Managing Chinese Outward Foreign Direct Investment

Managing Chinese Outward Foreign Direct Investment

From Entry Strategy to Sustainable Development in Australia

Xueli Huang
RMIT University, Australia

Ying Zhu
The University of South Australia

palgrave
macmillan

First published 2016 by
PALGRAVE MACMILLAN

Palgrave Macmillan in the UK is an imprint of Macmillan Publishers Limited, registered in England, company number 785998, of Houndmills, Basingstoke, Hampshire RG21 6XS.

Palgrave Macmillan in the US is a division of St Martin's Press LLC, 175 Fifth Avenue, New York, NY 10010.

Palgrave Macmillan is the global academic imprint of the above companies and has companies and representatives throughout the world.

ISBN 978–1–137–39458–3

This book is printed on paper suitable for recycling and made from fully managed and sustained forest sources. Logging, pulping and manufacturing processes are expected to conform to the environmental regulations of the country of origin.

A catalogue record for this book is available from the British Library.

Library of Congress Cataloging-in-Publication Data
Names: Huang, Xueli. | Zhu, Ying, 1961–
Title: Managing Chinese outward foreign direct investment : from entry strategy to sustainable development in Australia / Xueli Huang, Ying Zhu.
Description: New York : Palgrave Macmillan, 2016. | Includes index.
Identifiers: LCCN 2015026342 | ISBN 9781137394583 (hardback)
Subjects: LCSH: Investments, Chinese—Australia. | Corporate governance—China. | International business enterprises—Management. | Diversity in the workplace—China. | Corporations, Chinese. | BISAC : BUSINESS & ECONOMICS / Management. | BUSINESS & ECONOMICS / Strategic Planning. | BUSINESS & ECONOMICS / Industrial Management. | BUSINESS & ECONOMICS / Outsourcing.
Classification: LCC HG5892 .H83 2016 | DDC 338.8/8951094—dc23
LC record available at http://lccn.loc.gov/2015026342

Contents

Tables and Figures

Tables

Figures

Foreword

Chinese enterprises made their first investments in Australia less than three decades ago. The CITIC Group's investment in the Portland aluminium smelter and the China Metallurgical Import and Export Corporation's investment in the Channar iron ore project were pioneers of Chinese investment in the outside world as well as in Australia. The Chinese government chose to dip its toe in the water of direct investment outside greater China in Australia with these two projects in the resources sector, just as it had tested ideas and practice on technical assistance from abroad and regional and global trade diplomacy through interaction with Australia.

Chinese foreign direct investment abroad was tiny by global standards until the twenty-first century. It is now a mainstay of expansion of international economic relations. By 2013, annual flows of direct investment from China exceeded US$100 billion. Australia remains a major destination, accounting for more than from any other sovereign country other than the United States. The resources sector remains of central importance, accounting for almost two-thirds of the total stock of Chinese direct investment in Australia by 2013. This recent investment has been characterised by much greater sectoral diversity.

Chinese outward foreign direct investment abroad is on the way to becoming an even larger element of international economic cooperation. Australia's high complementarity with the Chinese economy, the close bilateral economic ties between the two countries and the increasing scale of the Chinese economy ensure that Chinese investment in Australia will become more important to the development of the two countries. A close study of China's direct investment in Australia can reveal important realities that are important not only for Chinese and Australian development, but also in global terms.

In this book, Xueli Huang and Ying Zhu take a thorough look at the experience of Chinese direct investment in Australia. Their conclusions are instructive for future management of Chinese direct investment abroad and for management of incoming investment by Australia and other recipient countries.

The book is comprehensive in its scope. It is grounded on familiarity with and sound application of the international literature on foreign direct investment, and each chapter goes back to these foundations.

It draws on a database that includes the major examples of Chinese investment in Australia since the beginnings. The analysis of average results from the data set reveals general tendencies, and well-chosen case studies allow the authors to dig deeper to obtain qualitative insights into particular issues. The case study of the Sino Iron Project becomes a source of a long list of errors to avoid. However, the success of the Minmetals' acquisition of all but one of the mines of Oz Minerals in the trough immediately following the Great Crash of 2008 illustrates what can go right.

The book examines the reasons for investment, as well as the determinants of success. The financial success rates are low, underlying the importance of analysis of reasons for generally poor performance. It turns out that one crucial determinant of success with resource investment is timing in the business cycle. Don't invest at the peak of the cycle. But how do you know where you are?

Chapters on integration of businesses acquired by Chinese firms, corporate governance, cross-cultural relations within Chinese enterprises, and sustainable development and community relations take us into more subtle determinants of success. These parts of the book are of interest to students of International Relations and cross-cultural relations beyond the business sector. The main lessons are cautionary. The authors argue that Chinese business is on a long march to understanding the international environments that are increasingly important to its success.

Managing Chinese Outward Foreign Direct Investment: From Entry Strategy to Sustainable Development in Australia is a commendable contribution to understanding an increasingly important dimension of China's relations with Australia and the world. Future analysts of these issues will build from sound foundations laid down in this book.

Ross Garnaut
Professor of Economics
The University of Melbourne, Australia

Abbreviations

ASX	Australian Securities Exchange
CCC	cross-cultural competence
CCE	corporate community engagement
CCP	Chinese Communist Party
CEO	chief executive officer
CP	CITIC Pacific
CPM	CITIC Pacific Mining
CRM	community relations manager
CSA	country-specific advantage
CSP	corporate social performance
CSR	corporate social responsibility
CSRC	China Securities Regulatory Commission
FDI	foreign direct investment
FIRB	Foreign Investment Review Board
FSA	firm-specific advantage
GFC	global financial crisis
HCN	host-country national
HRM	human resource management
IB	international business
ICMM	International Council on Mining and Metals
IPO	initial public offering
JV	joint venture
KPI	key performance indicator
LOF	liability of foreignness
M&A	merger and acquisition
MCA	Minerals Council of Australia
MCC	Metallurgical Corporation of China
MMG	Minerals and Metals Group
MNC	multinational corporation
NDRC	National Development and Reform Commission
NGO	non-governmental organisation
OECD	Organisation for Economic Co-operation and Development
OFDI	outward foreign direct investment
OLI	ownership–location–internalisation
PCN	parent-country national

POE	privately owned enterprise
RBV	resource-based view
ROI	return on investment
SASAC	State-owned Asset Supervision and Administration Commission of China
SIP	Sino Iron Project
SME	small and medium-sized enterprise
SOE	state-owned enterprise
TCE	transaction cost economics
WOS	wholly owned subsidiary

1
Introduction

Chinese outward foreign direct investment (OFDI) has rapidly increased over the last two decades, from about US$1 billion in 1991 to US$107.8 billion in 2013 (Ministry of Commerce, 2014). Consequently, China has become an important source of global capital. Since 2008, China has been among the top ten countries globally in terms of OFDI (UNCTAD, 2014), ranked as tenth in 2008, fourth in 2009, third in 2010, seventh in 2011 and third in 2012 and 2013. In 2013, Chinese OFDI stock reached US$660.4 billion. However, it only accounted for about 2 per cent of the global FDI stock (UNCTAD, 2013). The Chinese government has indicated that its OFDI will be more than US$500 billion by the end of its 12th five-year plan (2011–2015). It is widely expected that in the next decade Chinese OFDI will exceed 1 trillion US dollars.

The major driving forces behind the rapid growth of Chinese OFDI at the macro-level have been the rise of the Chinese economy and the strong support from the Chinese government (Luo et al., 2010). Since 2011, the Chinese economy has become the world's second largest economy, 32 years of rapid economic development after China implemented its "open-door and economic reform" policy in 1979. Chinese government has supported its domestic firms to expand overseas by providing financial support and low-interest loans (Buckley et al., 2007), simplifying its OFDI approval process and improving its OFDI-related systems (Luo et al., 2010).

The rapid growth of Chinese OFDI over the last decade has been accompanied by a change in the international investment environment. Governments in many host countries have revised or developed new foreign investment policies (Economou & Sauvant, 2012), making FDI more challenging. Moreover, many multiple or bilateral agreements have been established globally, often in favour of FDI from the countries that are party to the agreements while discriminating FDI from other countries. At the societal level, communities in host countries increasingly expect and demand foreign multinational corporations (MNCs) to be more socially and environmentally responsible for their operations in host countries (Klein, 2014).

Many changes have also taken place in the domestic environment in Chinese OFDI. With the new generation of Chinese leadership elected in March 2013, and the slowdown of global economic recovery, the Chinese government has refined its economic policies that stress the importance of domestic consumption, innovation and change of industry structure from labour-intensive to high value-added in developing its national economy. Moreover, the Chinese government has put more emphasis on the performance of Chinese OFDI, particularly that of foreign subsidiaries of state-owned enterprises (SOEs).

This book focuses on the management of Chinese OFDI, particularly foreign subsidiaries established through merger and acquisition (M&A) at the organisational level. As shown in Figure 1.1, the amount of M&A, similar to the trend of overall Chinese OFDI growth, has substantially increased over the past ten years.

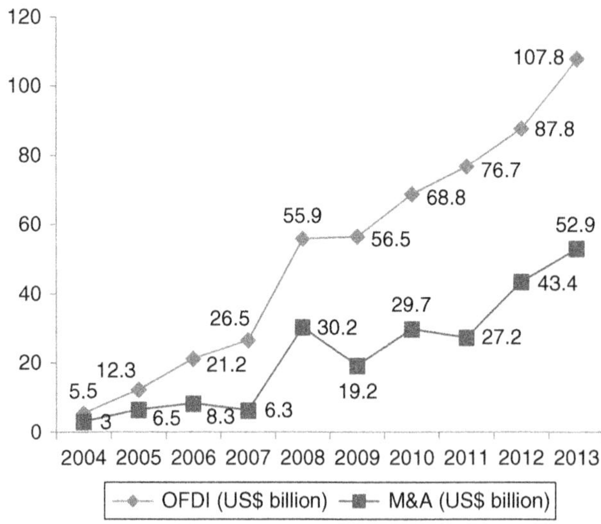

Figure 1.1 China's OFDI flow and M&A from 2004 to 2013
Source: Ministry of Commerce (2014).

The rise of Chinese OFDI has resulted in the establishment of 25,400 foreign subsidiaries owned by more than 15,000 Chinese firms in 184 countries or regions by the end of 2013 (Ministry of Commerce, 2014). The number of Chinese firms setting up their foreign subsidiaries has been increased from less than 7,000 in 2007 to 15,300 in 2013, more than doubling in six years (Ministry of Commerce, 2008, 2014). The Chinese SOEs have been responsible for a majority of Chinese OFDI so far, although their investment proportion was reduced from 81 per cent in 2006 to 55.2 per cent in 2013 due to the increase of OFDI from Chinese private firms (Ministry of Commerce, 2014).

How do Chinese MNCs manage their OFDI, particularly their foreign subsidiaries to achieve their investment objectives? This is a very important question as the role played by Chinese OFDI has increasingly become important to Chinese investors. OFDI is a complex, multidisciplinary and dynamic process, and it covers a wide range of issues in strategic management, finance and organisational behaviour. The exponential growth of Chinese OFDI over the last decade has substantially expanded the geographic scope of Chinese MNC's operations, enhanced their capability to manage international business and gained many financial and non-financial benefits. Nevertheless, Chinese MNCs have encountered many challenges in not only making their OFDI decisions but also subsequently managing these OFDIs after the transactions in order to ensure sustainable development. Anecdotal evidence has shown that the financial performance of Chinese OFDI has not met the expectation of Chinese investors, including the State-owned Asset Supervision and Administration Commission of China (SASAC) which owns the SOEs. As a result, the SASAC issued two interim regulations in 2011 for tightening the management and supervision of the overseas assets owned by Chinese SOEs. In particular, this move targeted centrally controlled SOEs, aiming to improve the OFDI decision-making process and subsequent corporate governance and management of the foreign subsidiaries to enhance their financial performance.

Although much research has been devoted to Chinese OFDI over the last decade, a recent literature review (Deng, 2012) found that "research to date has been fragmented and piecemeal, and numerous theoretical and empirical areas of the internationalisation of Chinese firms (ICFs) remain significantly under investigated" (p. 423). OFDIs, particularly those resulting in the establishment of foreign subsidiaries through M&A, should be regarded as a process that involves pre-acquisition preparation, acquisition, post-acquisition integration and subsequent management. Each of these stages in the process can have substantial impact on the OFDI's financial performance. Extant research on Chinese OFDI so far is not only fragmented, but also lopsided with regard to the issues at the pre-acquisition stage, such as locational choices (Buckley et al., 2007; Kang & Jiang, 2012; Kolstad & Wiig, 2012), role of Chinese government (Gao et al., 2015; Luo et al., 2010; Voss et al., 2010), entry mode consideration (Cui & Jiang, 2009; Rui & Yip, 2008), motivation (Deng, 2009; Globerman & Shapiro, 2009) and ownership of Chinese firms (Cui & Jiang, 2012; Hong & Sun, 2006; Huang & Chi, 2014). Scant research has been devoted to the post-acquisition integration and its subsequent management.

This book is specifically concerned with key issues in managing foreign subsidiaries established by Chinese investors either through Greenfield investment or acquisition in Australia. We have focused our research on the managerial issues of Chinese OFDI since 2007 and have interviewed more than 100 senior managers working in many Chinese firms in Australia as part of several research projects in Chinese OFDI-related areas. The book is built on the valuable insights we gained from these research projects, and it addresses the managerial issues from strategic investment decisions to corporate sustainable development. Particular emphases have been placed on the post-acquisition integration and management, concentrating on major managerial issues such as liability of foreignness (LOF) mitigation, post-acquisition integration, corporate control and governance, human resources and cross-cultural management and corporate social responsibility (CSR).

Before we explain the structure of this book, we would like to briefly discuss Chinese OFDI in Australia, China's bilateral trade with Australia and the recent development of Chinese OFDI in Australia.

Chinese OFDI in Australia

Australia provides an excellent opportunity for studying the management of Chinese OFDI, as it has been one of the largest destinations for Chinese OFDI so far. Australia became the fifth largest destination for Chinese OFDI by the end of 2013, only trailing behind Hong Kong, the Virgin Islands, the Cayman Islands and the United States (Ministry of Commerce, 2014). The Chinese OFDI stock in Australia had reached US$17.45 billion, with nearly two-thirds (65.3 per cent) invested in the extractive industry by 2013 (Ministry of Commerce, 2014). This number has not considered investments made by Chinese MNCs through their foreign subsidiaries, particularly those from Hong Kong, such as CITIC Pacific, owned by CITIC Group, and Yangcoal, owed by Yang Kuang Co.

Both firms were listed in Hong Kong when they made their investments in Australia. More than 80 per cent of Chinese OFDI in Australia was made by Chinese SOEs (KPMG & China Studies Centre, 2014b).

Although Australia became the fifth largest destination for Chinese OFDI by the end of 2013, Chinese OFDI stock in Australia accounted for only 2.9 per cent of the Australia's total FDI stock (Ministry of Commerce, 2014; UNCTAD, 2014). This is far less than the investments of the United States (26 per cent), the United Kingdom (17 per cent), Japan (6 per cent) and even Singapore (3 per cent) in Australia (Australian Bureau of Statistics, 2015).

Bilateral trade and China–Australia Free Trade Agreement

The figure of Chinese OFDI in Australia is dwarfed by that of its bilateral trade with Australia. China has been Australia's largest trading partner since 2011. In the Australian financial year of 2013–2014, Australian exports to China reached A$107.5 billion and accounted for about one-third (32.9 per cent) of its total export, while its imports from China were A$52.1 billion, sharing 15.4 per cent of its total imports (Department of Foreign Affairs and Trade, 2015b). While iron ore and coal were ranked as the largest Australian exports to China (A$52.7 billion and A$9.1 billion, respectively), telecommunication equipment and parts and computers were the two largest imports by Australia from China (Department of Foreign Affairs and Trade, 2015a).

Besides minerals exports to China, the exports of Australia's agricultural products, or "soft commodities", such as wheat, red meat (e.g. beef and lamb), seafood, wine and dairy products, have also substantially increased over the last decade. For example, beef exports to China trebled over the past five years and reached over 120,000 tons in 2014, but this only accounted for 10 per cent of total Australian beef exports and 3 per cent of the total Chinese market share (Meat & Livestock Australia, 2015). The exports of Australian dairy products (liquid milk, milk powder, cheese and butter) to China have nearly doubled since 2008 (Dairy Australia, 2015).

China's economic rise increases not only its demand for Australia's commodities and agricultural products, but also its service exports, particularly education and tourism, to China. Education is Australia's third largest export industry. China is Australia's largest international student market. By the end of March 2015, there were more than 115,000 Chinese students studying in Australia, accounting for 27.9 per cent of the total number of international students in Australia (Department of Education and Training, 2015).

With regard to tourism, China was the second largest inbound tourist market for Australia in 2013 with more than 700,000 arrivals (Tourism Australia, 2014), increasing from less than 200,000 in 2003. Chinese tourists contributed A$4.8 billion to Australia's economy in 2013, becoming a major source for the Australian tourism industry. In its national strategic plan ("Tourism 2020"), Tourism Australia is expecting that China will be a major contributor to its growth in the future (Tourism Australia, 2011). The growth in bilateral trade between China and Australia will further result in the increase of Chinese OFDI in Australia.

With the signing of the China–Australia Free Trade Agreement on 17 November 2014 (Prime Minister of Australia, 2014), a wide range of bilateral trade and investment opportunities will be accessible. On the one hand, most import tariffs will be immediately or gradually eliminated by the Chinese government, particularly for agricultural products and processed food, such as dairy products, beef, live animal exports, seafood, wine, honey and fruit juice (Department of Foreign Affairs and Trade, 2015c). China will also open its market for Australia's services industry, such as financial, insurance, healthcare, legal, education, and tourism and travel-related services. On the other hand, Australia will lift its threshold for screening Chinese investment. Chinese private entities can invest up to A$1.094 billion without the need for approval from the Australian government's Foreign Investment Review Board (FIRB). However, investment by Chinese SOEs in Australia still needs approval from the FIRB, no matter how it is invested. Moreover, the investment screen threshold is lower for sensitive sectors, such as media, telecommunications and defence-related industries (Department of Foreign Affairs and Trade, 2015c) and can be subject to change by the FIRB. For example, the investment screen thresholds for foreign investors in agricultural land and agribusinesses were reduced from A$252 million to A$15 million and A$53 million, respectively, since 1 March 2015 (Foreign Investment Review Board, 2015), partly due to negative sentiment from the Australian public towards the acquisition of agricultural land by foreigners.

Recent development of Chinese investment in Australia

Chinese investment in Australia has gradually diversified from the mining industry since 2013, partly due to the slump in commodity prices. For example, the commodity metal index and iron ore price declined from A$202 and A$150 per ton in January 2013 to A$134 and A$57 in March 2015, respectively (Index Mundi, 2015). Nevertheless, Chinese firms continue investing in the Australian mining industry, although at a slower pace. An example of such investment was Baosteel's acquisition of Aquila Resources in partnership with Aurizon, an Australian listed company operating in railway and port transportation, in May 2014 for A$1.4 billion.

Chinese investment in Australia's infrastructure, real estate and agribusiness has been growing over the past several years. China State Grid acquired 19.9 per cent of the electricity supplier SP AusNet and 60 per cent of energy infrastructure company SPI Australia for A$3.7 billion in 2013 (KPMG & China Studies Centre, 2014b). China Merchants Group secured its 98-year lease of Port of Newcastle for A$1.75 billion in a joint venture with a local company (50:50) in 2014 (Port of Newcastle, 2015). Chinese firms have also started investing in agriculture. Major Chinese investment projects in the Australian agriculture industry include the acquisition of Tully Sugar mill by China's Oil and Food Company for A$136 million; Shangdong Ruyi's acquisition of a cotton farm, Cubbie Station, in Queensland in partnership with a local wool company (80:20) for A$300 million; and Shanghai Zhongfu's leasing of Ord River for 50 years to develop 15,200 hectares of irrigated agricultural land in Western Australia. Chinese private investors, including small and medium-sized enterprises (SMEs), have increasingly become active in investing in Australia, particularly in real estate and agriculture, such as vineyards, farmland or dairy production sites. Nevertheless, Chinese investment in Australian agriculture is still very small, just over A$1 billion and less than 2 per cent of their overall investment in Australia by the end of 2013 (KPMG & China Studies Centre, 2013). Given the vast opportunities in the improvement of infrastructure and water systems in the Australian agricultural industry in order to reduce supply chain costs and to increase its irrigated farmland (Australian Government, 2014), it is estimated that the Australian agricultural industry needs A$600 billion of investment by 2050 for improving farm productivity and increasing production (Port Jackson Partners, 2012).

More recently, Chinese firms have also considered Australia as a platform for raising capital through initial public offering (IPO). Given the transparency and simplicity of the IPO procedure in Australia, it is expected that more Chinese firms will come to Australia to raise money. Nevertheless, the financial performance of these companies so far is generally unsatisfactory partly due to the problems in corporate governance (Shanahan, 2014) and partly due to problems in their business model and operations.

Australia is well positioned to attract FDI and has been heavily relying on FDI for economic development since late 1900s (Caves

& Krause, 1984). By the end of 2013, Australia's FDI stock reached US$591 billion (UNCTAD, 2014), with the United States and the United Kingdom accounting for nearly 50 per cent. There are several major benefits for investing in Australia. For example, Australia is abundant in mineral resources and is geographically close to China. It also has a stable political system and a well-regulated and transparent business environment, as well as a highly educated and skilled workforce. From an industry perspective, the Australian mining industry is well advanced globally in terms of technology and management (Huang & Austin, 2011). Its biotechnology and medical technology industries are also leading the world. The Australian agricultural industry is renowned for its safety, quality and cleanness. The Australian services sector is well developed and a major contributor to its GDP.

As is the case in any other country, there are several challenges in doing business in Australia (KPMG & China Studies Centre, 2014a). These challenges include (1) costs of doing business in Australia, (2) infrastructure bottleneck, (3) institutional integration, (4) negative media coverage and (5) entry and approval process. There are also cultural and institutional differences between Australia and China (Huang & Austin, 2011).

The structure of this book

Based on the investment trend of Chinese OFDI in Australia, we have designed this book by focusing on the managerial issues of Chinese OFDI. By adopting a process perspective, our research has targeted key managerial issues from strategic entry considerations to corporate sustainable development.

Chapter 2 examines how entry considerations impact the financial performance of Chinese OFDI in the Australian mining industry. It focuses on three strategic entry factors: timing [before, during and after the global financial crisis (GFC), during the GFC and after the GFC], stage (exploration, development and production) of project invested and percentage of ownership (minor vs. major) held by Chinese investors. The chapter also describes the mainstream theory of FDI as well as the emerging theories that focus on the explanation of the behaviour of MNCs from transitional and developing countries. This is one of the first empirical studies on the financial performance of Chinese OFDI.

Chapter 3 investigates how the LOF of Chinese MNCs affects investment decisions and subsequent managerial behaviour. The Sino Iron

Project has been used as a case study. Sino Iron Project was invested originally by CITIC Pacific, a Hong Kong listed company, and then by CITIC Group, a large Chinese SOE, and it is the largest Chinese investment project (more than A$10 billion) in Australia so far. Coupled with its Greenfield entry mode, the case provides an excellent opportunity to identify, describe and analyse the LOFs encountered by Chinese MNCs in Australia, what strategies are used to mitigate these LOFs and how LOFs impact financial performance.

Chapter 4 concentrates on the post-acquisition integration. It first develops a conceptual framework of post-acquisition integration based on a literature review. Using a combination of secondary data and in-depth interviews with senior managers, the chapter looks at how post-acquisition integration was conducted by a Chinese SOE, Minmetals, after it acquired a majority of the assets of OZ Minerals, an Australian-listed mining company, for A$1.4 billion in 2009. We focus on several key issues in the post-acquisition integration, including integration strategy, corporate control, functional integration, cultural integration, and knowledge transfer and learning. The smooth and successful post-acquisition integration of this case has offered many managerial insights into the post-integration process by Chinese MNCs.

Chapter 5 describes the corporate control by Chinese MNCs over their subsidiaries in Australia. We focus on the appointment of board of directors and chief executive officers (CEOs) in the Chinese-controlled and -listed companies in the Australian Securities Exchange (ASX). We examine the CEOs and a number of Chinese insider directors appointed, their educational background and professional experiences and explore the reasons behind such appointments. This chapter also presents a brief developmental history of corporate governance in China and, using an institutional logics perspective, how such a historical development, together with the influence of China's institution and culture, impacts corporate control by Chinese MNCs in their foreign subsidiaries (Thornton et al., 2012).

Chapter 6 describes and discusses cross-cultural management and human resource management of Chinese subsidiaries operating in Australia. Using a case study method, the chapter focuses on identifying problems, developing adequate policies and implementing them in a culturally appropriate manner. Chinese firms in Australia confront new challenges, particularly in the areas of a new labour market environment and relevant labour laws and regulations. It is crucial for them to be familiar with such new institutional and economic environments and, at the same time, be able to adopt and adapt new ways of doing business and managing their workforce. This chapter could be useful for other companies to learn both positive and negative lessons, particularly for companies from transition economies operating in developed economies.

Chapter 7 explores CSR practices implemented by Chinese-controlled entities in Australia. CSR has become an important issue for MNCs, particularly for those operating in foreign countries. Specifically, the chapter examines the community engagement by a Chinese-controlled firm in the Australian mining industry. The chapter starts with a brief review of literature on CSR and constructs a framework of factors influencing CSR performance. It then describes the community engagement strategy and approach implemented by the Chinese-controlled firm and its motivations for undertaking such community engagement practices.

In conclusion, Chapter 8 highlights the research findings and discusses many ongoing challenges for Chinese OFDI and MNCs operating abroad. In addition, it provides some implications for improving future development in the hope that the relevant decision-makers and business leaders both in China and abroad may learn from these lessons and be more successful in future OFDI and MNC operations.

References

Australian Bureau of Statistics. (2015). 5352.0 – International investment position, Australia: Supplementary statistics, 2014. Retrieved 9 May 2015, from http://www.abs.gov.au/AUSSTATS/abs@.nsf/DetailsPage/5352.02014?OpenDocument

Australian Government. (2014). Green Paper on developing Northern Australia. Retrieved 1 May 2015, from https://northernaustralia.dpmc.gov.au/sites/default/files/papers/Green_Paper.pdf

Buckley, P. J., Clegg, L. J., Cross, A. R., Liu, X., Voss, H., & Zheng, P. (2007). The determinants of Chinese outward foreign direct investment. *Journal of International Business Studies, 38*(4), 499–518.

Caves, R. E. & Krause, L. B. (1984). *The Australian economy*. Sydney: George Allen & Unwin.

Cui, L. & Jiang, F. (2009). FDI entry mode choice of Chinese firms: A strategic behavior perspective. *Journal of World Business, 44*(4), 434–444.

Cui, L. & Jiang, F. (2012). State ownership effect on firms' FDI ownership decisions under institutional pressure: A study of Chinese outward-investing firms. *Journal of International Business Studies, 43*(3), 264–284.

Dairy Australia. (2015). Australian dairy industry in focus 2014. Retrieved 1 May 2015, from http://www.dairyaustralia.com.au/~/media/Documents/Stats%20and%20markets/In%20Focus/Australian%20Dairy%20Industry%20In%20Focus%202014.pdf

Deng, P. (2009). Why do Chinese firms tend to acquire strategic assets in international expansion? *Journal of World Business, 44*(1), 74–84.

Deng, P. (2012). The internationalization of Chinese firms: A critical review and future research. *International Journal of Management Reviews, 14*(4), 408–427. doi:10.1111/j.1468-2370.2011.00323.x

Department of Education and Training. (2015). Monthly summaries of international student enrolment data. Retrieved 1 May 2015, from https://internationaleducation.gov.au/research/international-student-data/pages/default.aspx

Department of Foreign Affairs and Trade. (2015a). Australia's top 10 two-way trading partners. Retrieved 1 May 2015, from http://dfat.gov.au/trade/resources/trade-at-a-glance/Pages/default.aspx

Department of Foreign Affairs and Trade. (2015b). Australia's trade in goods and services 2013–2014. Retrieved 1 May 2015, from http://dfat.gov.au/about-us/publications/trade-investment/australias-trade-in-goods-and-services/Pages/australias-trade-in-goods-and-services-2013–2014.aspx

Department of Foreign Affairs and Trade. (2015c). China-Australia free trade agreement: Key outcomes. Retrieved 1 March 2015, from http://dfat.gov.au/trade/agreements/chafta/fact-sheets/Pages/key-outcomes.aspx

Economou, P. & Sauvant, K. (2012). Recent trends and issues in foreign direct investment, 2010. In K. P. Sauvant (Ed.), *Yearbook on international investment law & policy 2010–2011* (pp. 3–36). New York: Oxford University Press.

Foreign Investment Review Board. (2015). Recent changes to policy. Retrieved 1 May 2015, from http://www.firb.gov.au/content/policy.asp?NavID=1

Gao, L., Liu, X., & Lioliou, E. (2015). A double-edged sword: The impact of institutions and political relations on the international market expansion of Chinese state-owned enterprises. *Journal of Chinese Economic and Business Studies, 13*(2), 105–125.

Globerman, S. & Shapiro, D. (2009). Economic and strategic considerations surrounding Chinese FDI in the United States. *Asia Pacific Journal of Management, 26*, 163–183.

Hong, E. & Sun, L. (2006). Dynamics of internationalization and outward investment: Chinese Corporations' Strategies. *The China Quarterly, 187*, 610–634.

Huang, X. & Austin, I. (2011). *Chinese investment in Australia: Unique insights from the mining industry*. New York: Palgrave-Macmillan.

Huang, X. & Chi, R. (2014). Chinese private firms' outward direct investment: Do a firm's ownership and size matter? *Thunderbird International Business Review, 56*(September/October), 393–406.

Index Mundi. (2015). Iron ore monthly price – US Dollars per dry metric ton. Retrieved 1 May 2015, from http://www.indexmundi.com/commodities/?commodity=iron-ore

Kang, Y. & Jiang, F. (2012). FDI location choice of Chinese multinationals in East and Southeast Asia. *Journal of World Business, 47*(1), 45–53.

Klein, N. (2014). *This changes everything: Capitalism vs the climate*. London, UK: Allen Lane.

Kolstad, I. & Wiig, A. (2012). What determines Chinese outward FDI? *Journal of World Business, 47*(1), 26–34.

KPMG & China Studies Centre. (2013). Demystifying Chinese investment in Australian agribusiness. Retrieved 1 January 2015, from http://www.kpmg.com/AU/en/IssuesAndInsights/ArticlesPublications/china-insights/Documents/demystifying-chinese-investment-australian-agribusiness-october-2013.pdf

KPMG & China Studies Centre. (2014a). Demystifying Chinese investment: Chinese investors in Australia survey 2014. Retrieved 1 May 2015, from http://demystifyingchina.com.au/reports/Demystifying-Chinese-Investment-Survey.pdf

KPMG & China Studies Centre. (2014b). Demystifying Chinese investment: March 2014 update. Retrieved 1 May 2014, from http://www.kpmg.com/AU/en/IssuesAndInsights/ArticlesPublications/china-insights/Documents/demystifying-chinese-investment-in-australia-march-2014.pdf

Luo, Y., Xue, Q., & Han, B. (2010). How emerging market governments promote outward FDI: Experience from China. *Journal of World Business, 45*(1), 68–79.

Meat & Livestock Australia. (2015). Market snapshot: China. Retrieved 1 May 2015, from http://www.mla.com.au/NLRSReportDownload/China-Beef-Market-Snapshot-04-May-2015.PDF

Ministry of Commerce (Producer). (2008). 2007 Statistical bulletin of China's outward foreign direct investment. Retrieved from http://fec.mofcom.gov.cn/tjzl/jwtz/39067.shtml

Ministry of Commerce. (2014). *2013 Statistical bulletin of China's outward foreign direct investment.* Beijing: China Statistics Press.

Port Jackson Partners. (2012). ANZ insight – Greener pastures: The global soft commodity opportunity for Australia and New Zealand.

Port of Newcastle. (2015). Shareholders. Retrieved 1 May 2015, from http://www.portofnewcastle.com.au/Company-Information/Shareholders.aspx

Prime Minister of Australia. (2014). Conclusion of China-Australia free trade agreement. Retrieved 1 March 2015, from http://www.pm.gov.au/media/2014-11-17/conclusion-china-australia-free-trade-agreement

Rui, H. & Yip, G. (2008). Foreign acquisitions by Chinese firms: A strategic intent perspective. *Journal of World Business, 43*(2), 213–226.

Shanahan, L. (11 June 2014). Watchdog probes Sino Australia oil and gas for market fraud. *The Australian.* Retrieved from http://www.theaustralian.com.au/business/watchdog-probes-sino-australia-oil-and-gas-for-market-fraud/story-e6frg8zx-1226949914174

Thornton, P., Ocasio, W., & Lounsbury, M. (2012). *The institutional logics perspective: A new approach to culture, structure, and process.* Oxford: Oxford University Press.

Tourism Australia. (2011). Tourism 2020. Retrieved 1 May 2015, from http://www.tourism.australia.com/statistics/tourism-2020.aspx

Tourism Australia. (2014). China market profile. Retrieved 1 May 2015, from http://www.tourism.australia.com/documents/Markets/MarketProfile_China_May14.pdf

UNCTAD. (2013). World investment report. Retrieved 1 July 2013, from http://www.unctad-docs.org/files/UNCTAD-WIR2013-Full-en.pdf

UNCTAD. (2014). World investment report. Retrieved 26 June 2014, from http://unctad.org/en/PublicationsLibrary/wir2014_en.pdf

Voss, H., Buckley, P. J., & Cross, A. R. (2010). The impact of home country institutional effects on the internationalization strategy of Chinese firms. *Multinational Business Review, 18*(3), 25–48.

2

Strategic Entry Considerations and Their Impact on Investment Performance in the Australian Mining Industry

Introduction

Despite the rapid increase of OFDI from China and other developing countries and emerging economies, anecdotal evidence (Liu & Waldemar, 2011) has shown that the financial performance of these OFDIs varies substantially, and disturbingly, a large proportion of them fail. Although research over the last three decades has been dedicated to the theoretical and practical understanding of OFDI performance and its influencing factors, an overwhelming proportion of this research has focused on the OFDI from developed countries (Slangen & Hennart, 2008). Research on OFDI from developing countries to developed countries is growing and has so far concentrated on the role of institutions (Meyer, 2004), the firm's ownership (Cui & Jiang, 2012; Cui et al., 2011) and investment motives (Deng, 2009; Rui & Yip, 2008).

This chapter aims to explore the financial performance of Chinese OFDI and its influencing factors in the Australian mining industry. Specifically, this chapter attempts to examine the financial performance of major Chinese OFDI in the Australian mining industry, including those listed on the Australian Securities Exchange (ASX), and analyses how such performance is influenced by four entry factors: (1) market entry timing, (2) developmental stages of the target companies, (3) ownership level held by Chinese investors and (4) mode of entry (greenfield vs. acquisition). We employ multiple theoretical perspectives, such as the institution perspective, the resource-based view (RBV), transaction cost theory and agency theory, to explain why such entry behaviours are observed in Chinese investors.

Theoretical foundations

Mainstream FDI theory

Why do firms make FDIs? This has been a fundamental question in the study of FDI since the early 1960s. A seminal work can be traced back to Hymer (1976) who based his analysis of MNC's foreign operations on Bain's (1956) concept of oligopolistic advantages, including superior technology, economies of scale, knowledge and distribution channels. Firms must possess oligopolistic advantages to offset their disadvantages caused by their unfamiliarity with foreign countries. This view is broadly in line with the RBV (Barney, 1991) developed in the area of strategic management, which emphasises the important role an organisation's internal resources [e.g. "strategic capability" (Johnson et al., 2011 or "core competence" (Prahalad & Hamel, 1990)] play in surviving and competing successfully. Hymer (1976) elaborated that MNCs can exploit these advantages by establishing foreign operations. Essentially, Hymer's (1976) explanation of FDI focused solely on *internal*, or ownership, advantages that can be used to seek monopolistic rent for MNCs (Rugman, 1986).

Building on Coase's (1937) work on transaction costs, Buckley and Casson (1976) offered an alternative explanation of FDI. They argued that firms can reap the benefits of internalising missing and/or imperfect *external* activities in their value chains, thus reducing their transaction costs such as information, bargaining and enforcement costs. In contrast to Hymer's (1976) oligopolistic advantage, Buckley and Casson's (1976) internalisation theory is exogenous and focused on efficiency rent seeking. While Hymer's work is built on monopoly advantage, Buckley and Casson's argument is based on Coase's (1937) transaction costs. These two theories are similar to the concepts of "strategising" and "economising" (Williamson, 1991) used in the area of strategic management to explain how "economic rent" can be generated by firms.

Dunning (1977) has concentrated on the valuable input offered by different geographic locations (countries) in international production, including endowments such as natural resources, kind of labour, proximity to markets and technologies, and the legal and commercial environment such as market structure, government legislation and policies, to improve an organisation's performance and/or enhance its organisational capabilities. This is the locational advantage, which is often available to all firms operating in that location.

The most widely used theory for explaining international business activity and behaviour is probably Dunning's eclectic theory and paradigm (Dunning, 2001), incorporating monopolistic advantage (Hymer, 1976) and internalisation theory (Buckley & Casson, 1976). The eclectic theory states that: (1) MNCs can possess some assets that can be leveraged in other countries, that is, the ownership advantage; (2) different countries can offer different opportunities for MNCs, be it market, efficiency, natural resources or strategic assets; this is the location consideration; and (3) MNCs select means, such as the market entry model, that are appropriate for *internalising* its international business activities based on their ownership and location considerations. This is the ownership–location–internalisation (OLI) model. Briefly, the eclectic theory aims to explain why MNCs, particularly those from developed countries, make FDIs in foreign countries and identifies "why", "how" and "where" FDIs are made. The theory also identifies the requirements for a firm's success abroad (Moon & Roehl, 2001).

Emerging FDI theories

Dunning's eclectic paradigm (2001) sheds much insight on why firms make international investments and explains effectively the phenomenon of FDI flow from developed countries to developing ones. However, it has been challenged by several researchers (Mathews, 2006; Moon & Roehl, 2001), primarily on the grounds that it cannot adequately account for FDI by MNCs from developing countries (Mathews, 2006; Moon & Roehl, 2001). Dunning's eclectic paradigm was primarily developed during the period 1960–1990 and was based on the phenomenon of MNCs from developed countries establishing their operations in foreign countries, particularly in developing countries. Nevertheless, emerging economies played a crucial role in FDIs and were regarded as "the new FDI powerhouses" by the UNCTAD (2011, p. xii) and as "reshaping the structure of international business" (Gammeltoft et al., 2010, p. 95).

Many MNCs from emerging economies, such as China, India and Brazil, have invested heavily in developed countries in order to acquire strategic assets and have not had firm-specific advantage (Li, 2003); both are not well covered by the elective paradigm. MNCs from these emerging and other developing economies, although diverse, have displayed some common characteristics that differ from their counterparts in developed countries (Gammeltoft et al., 2010). First, most of them do not possess ownership advantage, such as technology, management know-how and famous brand names that are exploited by their counterparts from developed countries (Buckley et al., 2007; Hymer, 1976) when investing in developed countries, such as Australia, the United

States and Europe. Second, the home government of emerging MNCs plays a different role in promoting OFDI (Luo et al., 2010). Third, these MNCs have different motivations. Rather than seeking rent from their monopolistic assets, their motivations are more likely to be strategic asset seeking (Deng, 2009; He & Lyles, 2008) and learning (Deng, 2010; Liu & Woywode, 2013). Fourth, they have different internationalisation patterns and paths, such as the "springboarding" approach used by Chinese MNCs (Luo & Tung, 2007). Finally, the starting points of internationalisation for MNCs from emerging economies are different as the MNCs are latecomers and are competing in an era of extensively well-connected and increasingly globalised economies (Gammeltoft et al., 2010). Take OFDI from Chinese MNCs, for example – at the macro-level, Chinese OFDI has been driven by capital market imperfections, ownership advantages (e.g. relational assets and familiarity of operating in emerging market context) and institutional factors, such as a high level of government support (Buckley et al., 2007; Luo et al., 2010). At the organisational level, Chinese firms may use their OFDI to achieve their strategic goals (Rui & Yip, 2008) and seek both strategic assets and markets (Buckley et al., 2008). They have also leveraged their business network ties in order to promote their OFDI (Yiu et al., 2007). Learning is another motivation for Chinese OFDI, particularly at the early stage of internationalisation (Teagarden & Cai, 2009). The aforesaid considerations indicate that OFDIs from emerging economies, such as China, could be very different from those undertaken by firms in developed countries.

Recognising the limitations of the existing mainstream FDI theories, researchers have developed several other theories aiming to explain FDI behaviour by MNCs from emerging economies. This includes the institution perspective (Buckley et al., 2008; Child & Rodrigues, 2005; Meyer, 2004), the Link, Leverage and Learning (LLL) framework (Mathews, 2006) and the imbalance theory (Moon & Roehl, 1993). These newly developed theories recognise the influence of institutions, particularly those in the home-country, on the FDI and business networks of MNCs and, more importantly, the acquisition of strategic assets as an important motive of FDI for MNCs from developing countries, to address their competitive disadvantages. For example, Mathews (2006) described how "the Dragon multinationals", a group of latecomer and newcomer MNCs from the Asia Pacific, accessed needed resources for their growth through their international connections. He argued that the world economy is more closely linked due to rapid globalisation; thus the inter-firm connections seen as value chain activities are spread geographically based on production factor considerations. These inter-firm connections, such as partnerships, alliances or acquisitions and external rich resource reservoirs (e.g. technology, knowledge and experience) abroad, are not otherwise available and encourage dragon multinationals to use these inter-firm linkages and leverage the resources accessed from their connections abroad. These companies learn through repeated applications of linkage and leverage – the so-called Linkage, Leverage and Learning (LLL) Framework (p. 20), to improve their competitive advantage and/or strategic position.

Other FDI theories

Besides the key FDI theories mentioned above, several other theories have also been developed examining FDI and its dynamics from the perspectives of product, industry and country. Built on the concept of product life cycle, Vernon (1966) argued that FDI in production abroad is made by the producer as a new product evolves from the introduction stage to maturing and standardised stages, particularly from developed countries to less developed countries, due to competitive pressure and costs. Adopting a macroeconomic approach to analysing FDI by Japanese MNCs and based on comparative advantage and costs, Kojima (1982, p. 2) proposed that "[direct] foreign investment should originate in the investing country's comparatively disadvantaged (or marginal) industry (or activity), which is potentially a comparative advantaged industry in the host country" to benefit both home and host countries. He also cited industries using natural resources, agricultural raw materials and foods, textiles, car assembly lines, electric equipment and appliances, as examples of Japanese MNCs making FDIs abroad due to comparatively high labour costs in Japan (Kojima, 1978). At the country level, Ozawa (1992) suggested that a country's OFDI progresses in stages; he outlined the pattern (nature and direction) of its changes according to its economic development. As the country's economy develops from a factor-driven to an investment-driven stage, it invests outward in labour-intensive manufacturing and resource extraction and then invests in intermediate goods industries as it moves from the investment-driven stage to the innovation-driven stage.

Another theoretical perspective that has been used to explain the international expansion behaviour of MNCs is the Uppsala internationalisation model (Johanson & Vahlne, 2009; Johanson & Wiedersheim-Paul, 1975; Vahlne & Johanson, 2013). This model considers a lack of knowledge as a major barrier to the expansion of international operations (Johanson & Vahlne, 2009; Johanson & Wiedersheim-Paul, 1975).

Learning through experience in the international market is necessary for increased knowledge about foreign markets as well as to overcome "liabilities of foreignness" (Zaheer, 1995) and "double layer acculturation" (Barkema et al., 1996). Learning through networking has become crucial to overcoming "liabilities of outsidership" (Johanson & Vahlne, 2009; Vahlne & Johanson, 2013), with business networks playing an increasingly important role in international operations. As firms increase their knowledge of foreign markets, they increase their commitment to international operations, which in turn helps with knowledge accumulation. Three important ways to gain knowledge are acquiring, developing and networking.

In summary, given its increasing importance to both the national and the global economy, and MNCs' performance, FDI has been examined from several different theoretical perspectives and at different levels. While the debate continues on whether new theories should be developed to better account for a variety of FDI, particularly by MNCs from less developed countries (Yiu et al., 2007), it is increasingly recognised (Child & Rodrigues, 2005; Meyer et al., 2009; Peng et al., 2008) that the national institutions have a considerable impact on OFDI and should be considered in examining OFDI activities and patterns (Dunning, 2006) to enhance our understanding of OFDI, particularly by firms from emerging economies, such as China.

Building on the theoretical perspectives discussed above, the following subsections elaborate on the concept of four major entry considerations into the resource sector (Huang & Austin, 2011) and develop hypotheses about their influence on the financial performance.

Hypothesis development

We focused on two major types of strategic considerations for the timing of an investment in the mineral industry: (1) early entry versus late entry; and (2) commodity price cycle. The potential benefits for early movers include those derived from pre-emptive, economic, technological and behavioural factors (De Castro & Chrisman, 1995). Early movers can seize strategic opportunities that only appear on rare occasions (Bowman & Hurry, 1993). They also have the opportunity to accumulate knowledge about the market and develop competencies to deal with the host-country government and local business partners (Luo, 1999). In the mineral industry, early movers may be exposed to more opportunities to invest in high quality mineral projects in terms of mineral contents and resource size.

However, there are also risks and uncertainties associated with the timing of investment, particularly FDI. Investment uncertainty has been treated from two perspectives: endogenous and exogenous. Two theoretical perspectives can be used to explain these types of uncertainty: transaction cost economics (TCE) (Williamson, 1981) and real options theory (Kayali, 2006; Krychowski & Quélin, 2010). From a TCE perspective, endogenous uncertainty is about the behavioural uncertainty of the firm and can be partly attributed to the acquiring firm's experience in its operations in a foreign country. Some effective ways for solving problems arising from uncertainty include information-gathering and learning (Casson, 1994; Luo, 1999). Early movers can outperform late entrants because they have accumulated more experience than the latter in terms of market expansion and asset turnover, while late entrants tend to have superior performance than their early counterparts in terms of risk reduction and accounting return (Luo, 1999). Moreover, firms can make initial small investments for establishing a platform for gathering information and further investment, or platform investment (Kogut & Chang, 1996). Thus, sequential entry can be an effective approach for improving investment performance (Zhang et al., 2007).

Exogenous uncertainty includes factors that come from the business environment, such as price of minerals, currency depreciation and government regulations (Rivoli & Salorio, 1996). This type of uncertainty can only be reduced with the passage of time as the environment factors causing it are beyond the investor's control and thus can be analysed with a real option approach. The reversibility and delayability of investment, along with the nature of uncertainty, are two critical factors considered when making investment decisions about "when" to invest (that is, whether to "invest now" or "wait-and-see").

Both endogenous and exogenous uncertainties are faced by Chinese firms when making OFDI in the Australian mining industry. The level of endogenous uncertainty, particularly in the areas of international operations and management, is generally high given the short history of Chinese firms in international operations because their OFDI is a relatively new growth strategy. Like other latecomers (Morck et al., 2008) in the global market, Chinese firms have used OFDI as an important method for learning (Mathews, 2006) and a springboard to catch up with their competitors, particularly those from developed countries (Luo & Tung, 2007).

Exogenous or environment uncertainty in mining industry investments has a unique characteristic: the cyclical nature of global commodity prices, often strongly associated with the international economic cycle. It is very difficult to predict accurately when prices are at the peak or trough, and how long each growth and decline phase in the cycle will last. The recent rise of emerging economies, such as the BRICS (Brazil, Russia, India, China and South Africa) countries, has further complicated such predictions. One crucial consequence of this price cycle is its impact on the timing of investment as it affects the return on investment (ROI).

A high commodity price at the time of investment generally raises the expectation of sellers about its profitability and produces a high level of expected ROI. If the price cycle is random and frequent and the price volatility is low, there is no point in delaying, or waiting-and-seeing. However, the current commodity price cycle is relatively long and volatile (Garnaut, 2012), presenting an option for delaying the investment. The commodity price slump caused by the global financial crisis (GFC) in 2008 may present a once-in-a-generation opportunity for investment. While the early movers encounter and explore more endogenous uncertainty (such as mining project development and after-acquisition integration), both early and late entrants have to deal with environment uncertainty, particularly global commodity and economic cycles. Given their relatively short history of international operations, most Chinese firms lack experience and capability in dealing with endogenous uncertainty. Their ability to deal with exogenous uncertainty becomes crucial in making investment decisions. Moreover, during dire economic conditions, such as the GFC, the assets of host-country organisations may be low or undervalued and thus can become more financially attractive to MNCs (Agrawal & Jaffe, 2003; Williams & Martinez, 2012). This is particularly pertinent to organisations with financial backing of the home-country government, such as Chinese SOEs (Luo et al., 2010). On the basis of the above discussion, we posit the following hypotheses:

> H1a: Investments made in the mining industry during the GFC performed better than those before and after the GFC.
>
> H1b: Investments made in the mining industry at the peak of an economic cycle (in this case, immediately before the GFC) underperformed compared with those made during and after the GFC.

The developmental stage of the target firms invested

The developmental stage of the target project invested is another entry consideration that can influence an investment's financial performance.

This is a unique entry consideration for investment in the mining industry, as different levels of development involve changing degrees of risk and uncertainty. Broadly speaking, the mining life cycle can be classified into four stages: exploration, developmental (mine construction), production (or mining operations) and mine closure and rehabilitation (Minerals Council of Australia, 2012). With the exception of the mine closure stage, investment can be made in each of the first three stages. The exploration stage primarily involves mineral discovery activities until the feasibility study is completed to establish the extent of resources and their reserve. Following the exploration stage is the developmental (construction) stage, where processes are undertaken to bring the mine to production. The third stage is production where major activities, such as mine operations, transportation and marketing, are undertaken. The typical tasks involved in each of these three stages are presented in Table 2.1.

Table 2.1 Typical major tasks involved in each stage of a mining project development

Stage of development	Typical major activities	Milestone of completion
Exploration	Prospecting and exploration Minerals discovery (e.g. drilling and testing) Resources definition Pre-feasibility study Feasibility study	Feasibility report
Development	Government approvals and licensing (e.g. environmental approval, buildings and land use) Community consultations and engagement Financial approval Infrastructure development (e.g. road, railway, port, power and water) Mine construction and contracting Major supplier contracting (e.g. project engineering, procurement, construction and management) Ordering key machinery and equipment for mining operations	Start of production
Production	Mining operations Transportation Marketing	Closure of the mine

The exploration stage is probably the simplest among these three stages as it has a narrow scope of activities, including drilling and testing. It includes a pre-feasibility study and ends with a feasibility study. The major risks involved in this stage include locating the minerals deposit, securing an exploration licence and assessing the economic viability from both geo-scientific, engineering and operational perspectives (Minerals Council of Australia, 2012).

Various "liabilities of foreignness" (Zaheer, 1995) emerge at the developmental stage, given the nature and variety of tasks involved, such as government approvals, community consultations, dealing with local business partners, host-country regulations in industry relations and human resource management, project management and securing finance (Minerals Council of Australia, 2012). Dealing with government departments for various approvals and community consultations involves a great deal of uncertainty and is often time-consuming. Government evaluation criteria are often context-dependent and politically oriented, and the local community's expectations and demands are culturally embedded. It is at this stage that the problems of information asymmetry between the foreign investors and local operators, and those arising from the "liability of foreignness", are most acute.

Major problems at the production stage include mining operations, marketing and coordination of functional tasks. This stage also requires dealing with many key stakeholders, such as production supplies, minerals transportation and distribution agencies, and local communities. However, there are a few tasks that require dealing with government departments for approvals, community consultations and construction contractor management, thus lowering the levels of uncertainty and risk. Moreover, production generates cash, thereby reducing the pressure for financing.

Based on the nature of activities involved, the levels of uncertainty and risk, and the degree of information asymmetry, partly due to the liability of foreignness of Chinese investors, in each of these three stages, the following hypotheses can be developed:

H2a: The financial performance of investments in the production stage of the mine is better than that of investments in the developmental stage.

H2b: The financial performance of the investments in the production stage of the mine is better than that of investments in the exploration stage.

Mode of market entry and financial performance

Three broad modes of entry can be used by Chinese firms into the Australian mining industry: wholly owned foreign subsidiaries (acquisitions or Greenfield), joint ventures (JVs) (equity or marketing) and shareholding (majority or minority). The selection of ownership level is very important in entering foreign markets as it is associated with the level of control and risk (Chari & Chang, 2009), both of which contribute to investment performance.

The RBV (Barney, 1991) and transaction cost theory (Buckley & Casson, 1976; Coase, 1937) have been used in explaining the mode of market entry or "ownership structure" (Gomes-Casseres, 1989). The basic assumption is that net economic benefits vary with entry modes as each of them involves different costs and benefits and are dependent on the investing firm's (or MNC's) resources and costs of implementation (Gomes-Casseres, 1989). The strategic capabilities for an MNC to operate in a foreign country include geographical experience, industry experience, privileged access to markets or inputs, marketing skills and technologies (Gomes-Casseres, 1989). Based on RBV, a new subsidiary (the so-called "Greenfield investment") is feasible if the investing firm has the resources and competences, or "strategic capability" (Johnson et al., 2011), to survive and compete successfully in a host country. Otherwise, it needs to consider other entry alternatives, depending on whether the lacking strategic capability can be internally developed or acquired.

Transaction cost theory has been used for analysing the preferred market entry mode. If the lacking strategic capability for entering a foreign country is available in the local market, a JV with local firms and acquisition are two options. Otherwise, internal development is the only option. The preferred option depends on its costs and the nature of the lacking capabilities.

JV could be a preferred option if a local firm has the critical capabilities required by the MNC and the cost of transferring critical capabilities from the MNC and local firm to the JV is lower than acquisition (ownership). Characteristics of capabilities that are difficult to transfer to a JV include: tacit knowledge, those knowledge and competences embodied in managerial routine, contingency-based and potential spillover that negatively affect the contributor (Rivoli & Salorio, 1996 for detailed refs). Additionally, firms will be more reluctant to form a JV if substantial investment specific to the JV is required (Williamson, 1981).

There is an implicit fundamental assumption in the above analysis that the investing MNC has the appropriate resources and competencies to enhance the competitiveness and financial performance of its foreign subsidiary (in the case of either JV or acquisition). Unfortunately, this assumption is often not supported by Chinese firms when they invest in the Australian mining industry. Most Chinese firms generally do not have the strategic capabilities, except for financial resources, for their operations in the Australian mining industry. There are two reasons for this. First, they lack experience in international operations and management, particularly in developed countries such as Australia, though large Chinese SOEs can have advantages in accessing cheap financial resources (Buckley et al., 2007) due to the strong support by Chinese banks and government (Luo et al., 2010). Second, the Australian mining industry is renowned for its advanced technologies, management and global network (Drysdale, 2011). In comparison, Chinese firms have little competitive advantage over their Australian counterparts in these three areas.

The problems brought about by a lack of knowledge in and experience of, operations in foreign countries can be overcome by learning (Johanson & Wiedersheim-Paul, 1975; Johnson et al., 2008; Kogut & Chang, 1996). Learning by making small investments (such as investment options or platforms) as a minority shareholder at an early stage of internationalisation (Kogut & Kulatilaka, 1994), is an investment strategy widely used by Japanese firms (Kogut & Chang, 1996). This strategy could also be adopted by Chinese investors.

Research has shown that most Chinese MNCs have relied on Australian management teams – in wholly owned and JV contexts – to run their Australian businesses, particularly in larger-scale projects and in production (Huang, 2011). Chinese investors have little involvement in the management of Australian mines in which they have invested as minority (< 20 per cent) shareholders, although they may have a seat on the board. Another reason for minimal engagement in corporate management and governance activities in the firms is the very different managing structures between Australia and China and information asymmetry (Kang & Kim, 2010). Based on the discussion above, a clear relationship may not exist between the ownership and investment performance in the Australian mining industry. Thus, we propose:

> H3: There is no relationship between ownership level and financial performance in the Australian entities in which Chinese firms invest.

A particular case of mode of entry is wholly owned subsidiary (WOS) that can be established through Greenfield or acquisition of 100 per cent of the target company. This is a unique point (100 per cent) in the ownership. A Greenfield investment usually results in the establishment of a WOS and will require the investor to go through the whole developmental process, including exploration, development (construction) and production in the mining industry. While a 100 per cent acquisition can be made, strictly speaking, into the one of the three stages, we narrowly focused on those investments onto those mining companies or projects at the production stage. Entering a foreign country with the mode of WOS is regarded as being the most risky, compared with JV and shareholding. As discussed above, Greenfield investment can be more risky compared with acquisition of the entire target company at the production stage as it has to develop the mining project from the exploration and also require a variety of complex tasks in constructing the mining project and infrastructure (e.g. road, power, water and electricity) in order to bring the project into production. From the LOF perspective, acquisition also presents opportunities for the acquiring MNCs to retain the management team, thus alleviating the LOF in the production stage. The management team in the target firm could be very valuable to MNCs, particularly those from the emerging and developing countries as those MNCs, particularly those from China, often lack experience in international operations and business networks in host countries. Thus, we have the following hypothesis:

> H4: With WOS as the entry mode, MNCs entering with acquisition of the entire target firm at the production stage into the mining industry outperform those with Greenfield investment.

Methods

The database

Two criteria have been used to construct a database of Chinese investment in the Australian mining industry. First, the investment transaction must be more than A$10 million as this amount can be regarded as a major commitment. The second criterion is that the investments represent "10% or more of the voting power" of the invested entity to be consistent with the definition of FDI (OECD, 2008, p. 48). A database was compiled for these major Chinese investments in the Australian mining industry since 1987, which was when the first Chinese investment was made in a JV (Channar Project) between Rio Tinto and China's Metallurgical Import and Export Company (now Sinosteel Corporation). The database contains 27 Chinese firms with 33 investments in the Australian mining industry between 1987 and 2010.

Sample

We used both quantitative and qualitative approaches to collecting our data. Quantitatively, we used three criteria. First, the company in which Chinese firms invested had to be listed on the Australian Securities Exchange (ASX) and remain listed after receiving Chinese investment. This was to enable the stock-based market value of the investments to be calculated based on the stock price of the invested company. Second, the three marketing JVs (Channar JV between Rio Tinto and Sinosteel, the Bao-HI between Baosteel and Rio Tinto and the Wheelara JV between BHP and four Chinese steel mills) were excluded from this study as these three JVs are only a very small proportion of the operations of Rio Tinto or BHP. Their performance cannot be estimated accurately from the share prices of the companies in which they have invested. Third, only investments made before January 2010 were included as we believe a minimum of two years – although still short – should be allowed in order to assess the financial performance of investments. A total of 16 Australian companies met all these three criteria. These 16 cases were used to test hypotheses 1–3.

A problem with the data collection from the ASX was that those entering into the Australian mining industry with the WOS mode were not listed, or delisted after the acquisition. Thus, we could only collect qualitative data from media reports and the company's annual report. A total of five companies in the database constructed entered into the Australian Mining industry with WOS. Among them, four were through Greenfiled investment, while one through acquisition with the whole management team retained. These five companies were used to qualitative examine the influence of the entry mode (Greenfield vs. acquisition) on the financial performance.

Dependent variable

Several measurements have been widely used in assessing the FDI performance in extant literature. These include financial performance (such as level of profitability [Pan & Chi, 1999], sales level and sales growth [Brouthers, 2002]) and non-financial performance (such as market share, market access, marketing and reputation [Brouthers, 2002]). The objective financial performance data is more valuable for this study than the subjective data because the latter may suffer from management bias and inaccuracies. Financial performance is also one of the most popular measures in M&A (mergers and acquisitions) literature (Zollo & Meier, 2008), although "there is little or no agreement both across and within the disciplines on how to measure performance" (Zollo & Meier, 2008, p. 55). However, the financial performance measure may be difficult to obtain as firms are reluctant to provide such information.

This research uses a stock-based measurement as the dependent variable for Chinese investments as a proxy of financial performance of Chinese OFDI. This is an effective and simple measurement that calculates the differences between the share price invested and the share price at the end of 2011. The original number of transactions (number of shares acquired and their acquisition price) was traced from the target companies' announcements and the market value of the investments at the end of 2011 was calculated based on the share prices and the number of shares held by Chinese investors. The difference between the market value at the end of 2011 and initial investment was considered as the overall profit of the investment. The annual compound level of profitability was then examined by calculating the number of years the investment had been made. Where sequential investment was made at more than one point in time, a composite number of years was also calculated.

The financial performance of the five companies with the WOS entry mode was qualitatively estimated from their annual report and media.

Independent variables

Four independent variables were examined in this chapter: timing of market entry, stage of development, ownership level and the entry mode (Greenfield vs. acquisition). The timing of entry was categorised into three periods: between 2006 and September 2008 (when the GFC started), between October 2008 and June 2009 (when the global capital market started turning around) and July 2009 to early 2010. These three periods can be regarded as the three phases of an economic cycle: peak, trough and normal.

We classified the stage of development of a mining company based on the stage of its *major* mining project(s): exploration, development and production. These three stages of development are widely used in the mining industry to represent the progress of a mining project from exploration to production. We categorised the 16 companies studied into these stages based on the information collected from the company websites.

The percentage of ownership held by Chinese investors was gleaned from the target companies' websites or their annual reports at the end of 2011, particularly their announcements on the ASX. Greenfield and acquisition were used to examine the influence of entry mode on the financial performance.

Control variables

The value of the All Ordinaries Index was also collected from Yahoo Finance to benchmark the changes from the date of investment until the end of 2011. Where subsequent investments were made, a composite All Ordinaries Index was calculated. We also examined the impact of the exchange rate on investment performance by converting the Australian dollar value for the investment on the announcement date and the value on 31 December 2011 to Chinese currency (RMB) using the historical data on the currency converter at Yahoo Finance.

Findings and discussion

Overall patterns of financial performance

Table 2.2 shows several key characteristics of 16 Chinese investments in the Australian mining industry. To streamline the discussion, the Chinese investors are coded as C1 to C16 (Column 1). As shown in Table 2.2, the financial performance of Chinese investments varied substantially, with the overall return on investments ranging from –82. 7 to 90.4 per cent (Column 7). After taking into consideration the changes in the All Ordinaries Index, the overall profitability ranged from –64.3 per cent to 88.2 per cent (Column 8), with nearly two-thirds (11) of Chinese investors recording a loss.

The compound annual profit (Column 9), which varied from –40.4 per cent to 25.7 per cent is also shown in Table 2.2. Six Chinese investors were profitable with a compound annual profit ranging from 4.7 to 25.7 per cent. Ten Chinese investors, however, indicated a loss with seven of them having a compound annual profit of more than –20 per cent.

Timing of market entry and financial performance

The impact of market entry timing is shown in Figure 2.1. An inverted U-shaped curve can be observed, which corresponds to the three periods of the GFC: before GFC, during GFC and the market recovery period. The four investments made between January 2007 and August 2008 suffered heavy overall losses, ranging from –33.0 per cent to –82.7 per cent; the three investments made during the GFC performed very well with their overall profitability of 86.7 per cent, 16.7 per cent and 90.4 per cent. There was a flux of profit levels among the nine

Table 2.2 Characteristics of Chinese investment in the Australian mining industry

1	2	3	4	5	6	7	8	9	10
Chinese investor	**DoA[a]**	**Shares held (%)**	**Stage[f]**	**AORD on DoA**	**Change of AORD (%)[b]**	**Overall profit (%)**	**Diff'es[c]**	**Comp'd annual profit (%)**	**Comp'd annual profit (% RMB)[d]**
C1	March 2007	11.0	E	5,846.3	−29.7	−81.5	−51.8	−29.7	−29.1
C2	August 2008	11.4	E	5,039	−18.4	−82.7	−64.3	−40.4	−39.2
C3	September 2009	12.0	E	4,277.4	−3.9	41.7	45.6	17.4	22.2
C4	July 2009	13.0	E	4,048.3	1.5	12.0	10.5	4.7	11.2
C5	August 2009	15.0	E	4,464.4	−7.9	−10.0	−2.1	−4.4	0.2
C6	March 2009	17.4	P	3,121.7	31.7	90.4	58.7	25.7	44.76
C7	January 2010	19.9	D	4,889.8	−15.9	−36.0	−20.1	−24.5	−21.4
C8	August 2009	19.9	D	4,417.5	−6.9	−11.9	−5.0	−7	−10.2
C9	June 2009	19.9	D	4,047.3	1.6	−61.2	−62.8	−32.5	−34.2
C10	January 2007	36.3	D	4,250	−3.2	−33.0	−29.7	−18.1	−20.3
C11	November 2008	40.0	P	4,173	−1.5	86.7	88.2	21.8	19.5
C12	November 2009	41.3	E	4,277.4	−3.9	36.6	40.5	19.4	20.7
C13	December 2008	50.1	P	3,533.7	16.3	16.7	−0.39	6.4	15.8
C14	October 2009	55.0	D	4,575.2	−10.1	−58.5	−48.4	−33.3	−32.1
C15	September 2009	69.3	E	4,527.8	−9.2	−70.4	−61.2	−38.8	−36.6
C16	May 2008	56.0	E	6,035	−18.6	−75.9	−57.3	−32.5	−33.1

Note: a – Date of announcement; b – Change of All Ordinaries Index between the date of announcement and 31 December 2011; c – Difference between the overall profit and changes in AORD; d – Chinese currency: Ren Min Bi (RMB: Yuan); f – Developmental stage: E – exploration, D – development (construction), P – production; All Ordinaries Index at the end of December 2011 was 4111.

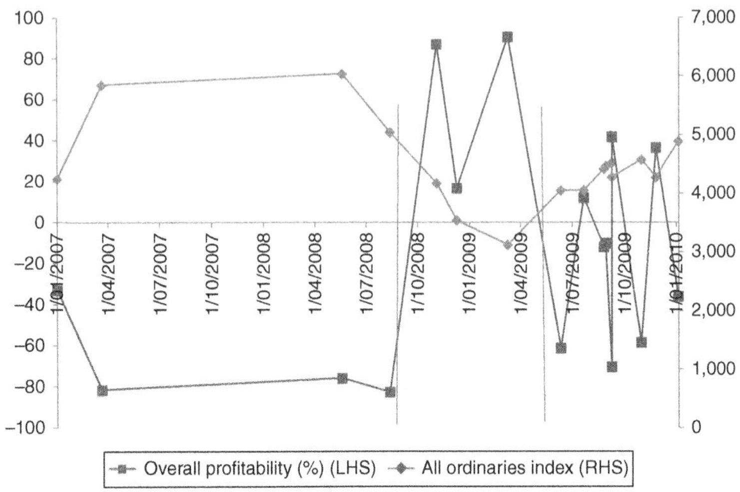

Figure 2.1 The relationship between entry time and the overall profit of Chinese investments

Chinese investments made after June 2009 when the situation on global financial market had started improving. The profit levels of these nine investments changed with three above the zero line and six under the line. All these findings have supported the first two hypotheses (H1a and H1b). It is also interesting to see that the window of opportunity for investment was only about eight months.

Stage of development invested and financial performance

Figure 2.2 shows the overall profitability and the compound annual profit for all 16 Chinese investments at the three different stages of the mining life cycle. We found that a U-shaped curve existed. A flux is evident for the eight investments in the exploration stage. All five mining companies in the developmental stage recorded a loss, while all three investments in production-stage companies were profitable.

Specifically, the eight companies at the exploration stage at the time of Chinese investment largely remained at the exploration stage by the end of 2011. The instability in financial performance at this stage needs closer scrutiny. Various reasons contributed to the positive performance of three mining companies at this stage. C12 is a Chinese-controlled entity and its Chinese parent specialises in mineral exploration. It appeared that the Chinese parent company had little intention of progressing to production and may have wished to

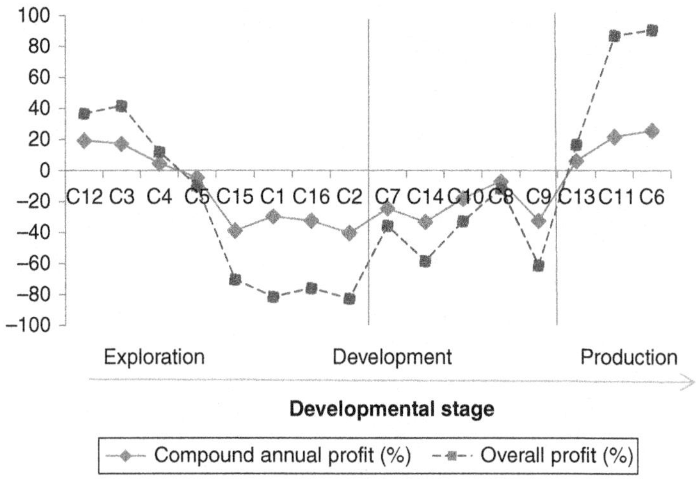

Figure 2.2 The financial performance of Chinese investments at different development stages of target firms

sell mineral assets discovered during exploration. Such a discovery-focused strategy can also be found in other Australian companies. C3 was acquired by an Australian mining company after the Chinese investment was made. Thus, the Chinese investor primarily benefited from the takeover premium (34 per cent in this case based on the bidder's statement). C4 was an investment made by a large Chinese steel mill at a relatively late stage (July 2009), representing an improvement on investment due to its learning experience.

All five Chinese investments in the mining companies at the developmental stage indicated a loss, with their compound annual profit varying from −7 per cent to −32.5 per cent. As shown in Table 2.1, the developmental stage typically requires a large amount of capital for infrastructure development, mining construction and equipment procurement. It is also the stage at which various government approvals have to be obtained and extensive dealings with local communities are needed; this involves a high level of uncertainty. The information asymmetry between Chinese investors and their target mining companies could also play an important role for investment in this stage as many Chinese investors lack international business experience and are unprepared for their investment in the global market.

All three investments in the Australian mining companies at the production stage were profitable. In fact, two of them were the best performers among all 16 Chinese investments in this study. Mining companies at the production stage have largely confirmed their minerals resources and reserves, obtained various government approvals and community support, developed the necessary infrastructure and established distribution networks. Moreover, they can generate cash flow from their own operations. Thus, the levels of risk and uncertainty are much lower than those at the exploration and developmental stages, although the quality of the minerals and production costs could still be concerns for investors.

The above findings also supported our hypotheses (H2a and H2b). That is, the financial performance of investments made during the mining companies' production stage is better than at the exploration or developmental stage.

Ownership percentage and financial performance

The ownership percentage held by Chinese investors in the companies and their financial performance is shown in Figure 2.3. An overall pattern of random variation around zero can be observed, with a declining treadline as the ownership percentage held by Chinese investors

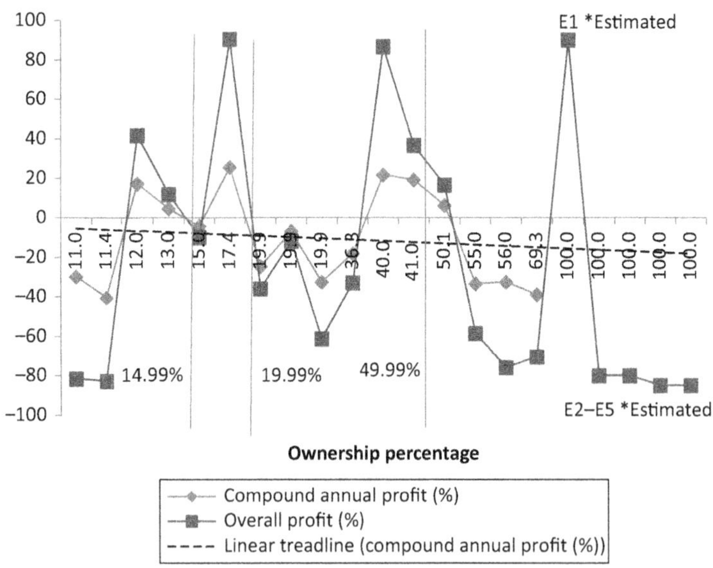

Figure 2.3 The relationship between ownership percentage and financial performance of Chinese investment

increased. We divided the ownership percentage into four categories: less than 15 per cent, 15–19.99 per cent, 20–49.99 per cent and 50 per cent or more, with five investments at 100 per cent. There is no clear pattern in any of these four sections in terms of gains and losses. However, it should be noted that three out of four of the companies where Chinese investors held more than 50 per cent ownership, indicated heavy losses (those five 100 per cent investments are separately discussed in the following section). This finding partially supports Hypothesis 3; that is, there is no relationship between the ownership percentage held by Chinese investors and the financial performance of the companies invested. Nevertheless, it seems that Chinese investment is more likely to perform poorly when the Chinese investor is a majority shareholder.

An further examination of the four mining companies in which Chinese investors hold a majority of shares reveals that C13 (50.1 per cent held by the Chinese investor) and C14 (55 per cent) have been relying primarily on their Australian team to govern and manage the company, while C16 (56 per cent) and C15 (69.3 per cent) have been governed and managed primarily by Chinese expatriates.

Mode of entry and financial performance

The impact of entry mode (Greenfield vs. acquisition) on the investment financial performance is also shown in Figure 2.3. Although all five investments resulted in the establishment of a WOS, their financial performance varied substantially. E1 was a large Chinese SOE and acquired a majority of assets from a listed Australian mining company. It was also able to retain the whole senior management team from the target company. This has been regarded as a crucial factor considered in its acquisition decision-making and was the only one successful investment among all five Chinese investments via the WOS entry mode. The timing of its acquisition also helped its financial performance as it was during the GFC. All other four Chinese investments with Greenfield investment had incurred heavy loss at the time of data collection. They all invested in magnetite, a type of low quality iron ore. E2 had developed its investment project through to production. However, its product cost was very high, together with substantial project delay and cost blowout. Other three Chinese investments (E3, E4 and E5) were not able to develop the project primarily due the lack of infrastructure (railway and port) for move the iron ore from the mine site for shipping. The finding in Figure 2.3 also qualitatively support our hypothesis (H4).

Discussion and conclusion

As presented above, a majority (14 out of 21) of Chinese investments in the Australian mining industry have underperformed (Table 2.2 and Figure 2.3). Among them, 10 out of 16 failed based on the benchmark of the All Ordinaries Index, although there was a large variation (Table 2.2). There are also clear patterns between performance and the timing of entry and the target company's developmental stage. These findings warrant discussion. First, it is well known that investments generally should not be made at the peak of economic cycles as assets are often expensive and overvalued. However, why did some Chinese firms – particularly Chinese SOEs – invest at such a time? This could be partly explained from an institutional perspective. Large numbers of loans from Chinese banks (with low interest rates) are readily available to Chinese SOEs due to their state ownership (Buckley et al., 2008). The perennial agency problem in Chinese SOEs due to state ownership and weak corporate governance (Liu, 2006; Tam & Yu, 2011) can also contribute to this situation. Using acquisitions for empire-building can be a dominant motive for SOE senior executives (Cooke, 2008; Morck et al., 2008). Another reason is management's "national pride", which suggests decision-making of individual decision-makers is influenced by national, social and political considerations and is often observed in the OFDI made by MNCs from developing countries in the developed countries (Hope et al., 2011). Moreover, extra cash is often available from internal operations only at the economic boom phase. All these can set the context for managers to make their OFDI even at the peak of the global economic cycle.

Three important practical implications can be suggested based on these findings. First, investment in the mining industry should avoid economic cycle peaks. Investing at such peak times is more likely to cause substantial loss than at other periods of the economic cycle. Moreover, the commodity price has also historically displayed a boom-to-bust cycle. With the current downturn of commodity prices, it surely worsened the return of investment made, but may represent opportunities for new investment. Second, the window of opportunity (in this case, the dramatic slump of the market during the GFC) for investment could be very short. This often requires quick investment decision-making. Therefore, "active-waiting" (Eisenhardt, 1999) can be very appropriate for investment in the mining industry, a process which demands a thorough analysis of potential target companies and advanced preparation of financial resources and government approvals. This disciplined and quick action is well demonstrated in the case of the Minmetals' acquisition of a majority of OZ Minerals' assets in 2009, which received the Best M&A Deal/Best Cross-Border M&A Deal Award in 2009 from *Finance Asia* magazine (*FinanceAsia*, 2009). Minmetals had been dealing with OZ Minerals well before its acquisition and conducted thorough analyses on OZ Minerals on two occasions before making an offer in January 2009, only 56 days after OZ Minerals was suspended from trading on the Australian Security Exchange. Minmetals amended the offer three days later after a request by the Australian Foreign Investment Review Board to exclude a copper mine of OZ Minerals from its previous acquisition offer. More importantly, Minmetals was also able to obtain the necessary approval (shareholders, banks and Chinese government) within two months. Another example of "active-waiting" is Baosteel, a Chinese state-owned enterprise (SOE), which acquired Aquila in July 2014. Baosteel made an investment of A$285.6 million in Aquila Resources, a junior Australian coal and iron ore developer, in August 2009 at A$6.50 per share. Baosteel made a second investment in December 2013 of A$ 2.10 and increased its shareholding to 19.8 per cent. It then launched a takeover of Aquila at A$3.40 per share in June 2014 when iron ore prices were down from A$140 to about A$90. The takeover was a joint bid (85–15 per cent) with Aurizon, the largest rail freight operator in Australia (Aurizon, n.d.).

The third implication is that investing as a minority shareholder in a listed mining company can be a viable strategy, particularly those at the production stage and if such investments can be reversed (sold) easily. The critical and challenging issue here is how knowledge can be gained by learning through such an investment strategy. This requires more emphasis on learning as part of an overall OFDI strategy. In this sense, JVs can be used as an effective investment approach.

The developmental stage of companies is crucial to investment performance. Our findings show that investments in target companies at the production stage can perform better than those made at the exploration and developmental stages, even if the investment criteria are likely to be higher than those used for the exploration or developmental stages. Furthermore, investments during the developmental stage resulted in losses. Crucial to the investment decision is access to information about the host government's approval procedures and regulations, target company's mining projects and management, and environmental protection and community engagement, which are often major challenges faced by

Chinese MNCs in developed countries like Australia (Sun et al., 2013). In this regard, experience and knowledge of international operations, particularly in the host country, are critical.

Many Chinese investors lack experience in international operations largely because of their long history of isolation from the global market. Although the Chinese government implemented the "open door" policy in 1979, OFDI for Chinese firms is a relatively new phenomenon and has only gained momentum since 2003 when Chinese privately owned enterprises were allowed to invest abroad (Zhang, 2009). Lacking information about the developmental process of a mining project in Australia due to the "liability of foreignness" (Zaheer, 1995) can contribute to the poor performance of investment in the exploration and, particularly, developmental stage because the latter usually involves a large variety of complex context-specific tasks and activities, as well as a large amount of capital as mine construction starts. The way of doing business in Australia differs significantly from that of China, and so does the cost of construction. Consequently, project-delays and cost overruns on mining projects are frequently reported in the Australian media, particularly at the developmental stage.

The disadvantage of being latecomers can also impact Chinese investment in the developmental stage of Australian mining projects. Like other latecomers in the Australian mining industry, Chinese investors face difficulties in making OFDI in companies with world-class mineral resources. The Australian mining industry has a long history and has been well developed. Global mining giants such as BHP Billiton and Rio Tinto have long conducted their explorations in Australia and secured the best mineral resources, such as the iron ore of the Pilbara region in Western Australia. These global giants normally do not need equity-based investment as they can raise money relatively easily on the global financial markets. Thus, Chinese investors have to focus their OFDI on emerging mining companies or those at the exploration and developmental stages.

We are not suggesting that investment in the exploration and developmental stages should not be made because they involve a wide range of uncertainties and a high level of risk. A major reason for such poor performance is the lack of knowledge about mining assets and their governance and management, as well as experience in doing business in Australia. A practical implication is to be better prepared before making investment decisions by gathering information and conducting due diligence. Another important implication is that a comprehensive view of the entire mining life cycle needs to be adopted to better understand the key uncertainties involved in each stage and thus better manage investment risks and improve investment performance.

Does the ownership level held by Chinese investors matter? Our findings indicate that it does not matter for Chinese MNCs, at least at the developmental stage. Rather, other factors, such as the quality of their board and management team, together with the invested or acquired mining assets, may play a crucial role in influencing financial performance. Many Chinese investors have relied primarily on an Australian team to manage the target companies (Huang, 2011), even though the Chinese company may have bought a majority of the shares (such as in the cases of C13 and C14) or even undertaken a 100-per cent takeover (as in the case of the Minmetals' acquisition of a majority asset of OZ Minerals). Again, lack of experience in international operations and scarcity of competent managers in Chinese investing firms can explain this situation. Due to the differences in corporate governance between Australia and China, Chinese investors play little role in corporate governance activities (Huang, 2011). Moreover, potential high transactional costs for post-acquisition integration – such as cultural clash and management turnover (Weber & Tarba, 2013) – can be another reason for Chinese investors not taking over the management of mining companies, even though they may hold a majority of the company's shares. It is the quality of the mineral assets of the target company and the Australian management team that have substantially influenced the financial performance of the investment. In the two cases (C15 and C16) where Chinese investors took over the management after acquiring a majority of the shares, the companies performed poorly. Both remain at the exploration stage. It is difficult to ascertain whether the Chinese parent company's "strategic capability" and management team, or the potential "governance discount" which international investors place on the Chinese-controlled entity, could contribute to its performance. Other important factors, such as entry timing and type of minerals, must also be considered.

Why, then, do Chinese investors want to control the companies in which they invest? This question could be partly answered from an institutional perspective. China has been a socialist country since the Communist Party of China took power in 1949, and it operated as a planned economy before the substantial economic reform of its national economic system at the beginning of this century. Under the planned economy, activities were conducted primarily through administration and bureaucracy, rather than an open market. This meant that the only way to control the assets was to own them. Such an embedded administrative heritage can still influence the thinking of Chinese business executives in making OFDIs. As discussed above, the availability of cheap bank loans to Chinese SOEs also helps them make large acquisitions. Culturally, Chinese have highly valued "trust" in business, which is primarily built on personal relationships (Chen, 2004). It may prove difficult for Chinese investors to trust someone they do not know to manage their investments. Moreover, corporate governance of Chinese MNCs is still weak (Tam & Yu, 2011). Good governance is particularly important to the success of MNCs from emerging economies (Khanna & Palepu, 2006).

These findings have raised an important question for Chinese investors: Is it important to control the company in which they have invested? From a financial perspective, our findings show that it is not necessary (or possibly desirable) to control the target company if the Chinese investing firms have little expertise in managing its operations. Rather, taking into consideration the quality of the board of directors and its senior management team of the target company should become a crucial part of pre-investment analysis and due diligence assessment. Thus, it may be preferable for Chinese investors to be minority shareholders if this does not affect their learning and knowledge accumulation. This not only reduces the level of their commitment, but also allows Chinese investors to be flexible in their subsequent investment decisions.

References

Agrawal, A. & Jaffe, J. F. (2003). Do takeover targets underperform? Evidence from operating and stock returns. *Journal of Financial and Quantitative Analysis, 38*(4), 721–746.

Aurizon. (n.d.). Corporate profile. Retrieved 1 January 2015, from http://www.aurizon.com.au/aboutus.

Bain, J. (1956). *Barriers to New Competition*. Cambridge, Mass: Harvard University Press.

Barkema, H. G., Bell, J. H. J., & Pennings, J. M. (1996). Foreign entry, cultural barriers, and learning. *Strategic Management Journal, 17*(2), 151–166.

Barney, J. B. (1991). Firm resources and sustained competitive advantages. *Journal of Management, 7*, 49–64.

Bowman, E. H. & Hurry, D. (1993). Strategy through the option lens: An integrated view of resource investments and the incremental-choice process. *Academy of Management. The Academy of Management Review, 18*(4), 760–782.

Brouthers, K. D. (2002). Institutional, cultural and transaction cost influences on entry mode choice and performance. *Journal of International Business Studies, 33*(2), 203–221.

Buckley, P. J. & Casson, M. (1976). *The future of the multinational enterprises*. London: Macmillan.

Buckley, P. J., Clegg, J. L., Adam, R. C., Hinrich, V., Rhodes, M., & Zheng, P. (2008). Explaining China's outward FDI: An institutional perspective. In K. P. Sauvant (Ed.), *The rise of transnational corporations from emerging markets: Threat or opportunity?* (pp. 715–748). Cheltenham: Edward Elgar.

Buckley, P. J., Clegg, L. J., Cross, A. R., Liu, X., Voss, H., & Zheng, P. (2007). The determinants of Chinese outward foreign direct investment. *Journal of International Business Studies, 38*(4), 499.

Buckley, P. J., Cross, A. R., Tan, H., Xin, L., & Voss, H. (2008). Historic and emergent trends in Chinese outward direct investment. *Management International Review, 48*(6), 715.

Casson, M. (1994). Internationalization as a learning process: A model of corporate growth and geographical diversification. In V. Balasubramanyam & D. Sapsford (Eds.), *The economics of international investment* (pp. 14–46). Aldershot: Edward Elgar.

Chari, M. D. & Chang, K. (2009). Determinants of the share of equity sought in cross-border acquisitions. *Journal of International Business Studies, 40*(8), 1277–1297.

Chen, X. (2004). On the intricacies of the Chinese Guanxi: A process model of Guanxi development. *Asia Pacific Journal of Management, 21*, 305–324.

Child, J. & Rodrigues, S. B. (2005). The internationalization of Chinese firms: A case for theoretical extension? *Management and Organization Review, 1*(3), 381–410.

Coase, R. (1937). The nature of the firm. *Econimoca, 4*(16), 386–405.

Cooke, F. L. (2008). *Competition, strategy and management in China*. New York: Palgrave Macmillan.

Cui, L. & Jiang, F. (2012). State ownership effect on firms' FDI ownership decisions under institutional pressure: A study of Chinese outward-investing firms. *Journal of International Business Studies, 43*(3), 264–284.

Cui, L., Jiang, F., & Stening, B. (2011). The entry-mode decision of Chinese outward FDI: Firm resources, industry conditions, and institutional forces. *Thunderbird International Business Review, 53*(4), 483–499.

De Castro, J. O. & Chrisman, J. J. (1995). Order of market entry, competitive strategy, and financial performance. *Journal of Business Research, 33*(2), 165–177.

Deng, P. (2009). Why do Chinese firms tend to acquire strategic assets in international expansion? *Journal of World Business, 44*(1), 74–84.

Deng, P. (2010). What determines performance of cross-border M&As by Chinese companies? An absorptive capacity perspective. *Thunderbird International Business Review, 52*(6), 509–524.

Drysdale, P. (2011). Foreword. In X. Huang & I. Austin (Eds.), *Chinese investment in Australia: Unique insights from the mining industry* (pp. xi–xiii). New York: Palgrave Macmillan.

Dunning, J. H. (1977). Trade location of economic activity and the multinational enterprise: A search for an eclectic approach. In B. Ohlin, P. O. Hesselborn, & P. J. Wiskman (Eds.), *The International Allocation of Economic Activity* (pp. 395–418). London: MacMillan.

Dunning, J. H. (2001). The eclectic (OLI) paradigm of international production: Past, present and future. *International Journal of the Economics of Business, 8*(2), 173–190.

Dunning, J. H. (2006). Comment on dragon multinationals: New players in the 21st century globalisation. *Asia Pacific Journal of Management, 23*(2), 139–141.

Eisenhardt, K. (1999). Strategy as strategic decision making. *Sloan Management Review, 40*(3), 65–72.

FinanceAsia. (2009). Awards for achievement 2009 – Best M&A deal/best cross border M&A deal: Minmetals' $1.2 billion acquisition of part of Oz Minerals. Retrieved 1 February 2012, from http://www.greenhillcaliburn.com/write/Finance%20Asia.pdf.

Gammeltoft, P., Barnard, H., & Madhok, A. (2010). Emerging multinationals, emerging theory: Macro- and micro-level perspective. *Journal of International Management, 16*, 95–101.

Garnaut, R. (2012). The contemporary China resources boom. *Australian Journal of Agricultural & Resource Economics, 56*(2), 222–243.

Gomes-Casseres, B. (1989). Ownership structures of foreign subsidiaries: Theory and evidence. *Journal of Economic Behavior & Organization, 11*(1), 1–25.

He, W. & Lyles, M. A. (2008). China's outward foreign direct investment. *Business Horizons, 51*(6), 485–491.

Hope, O.-K., Thomas, W., & Vyas, D. (2011). The cost of pride: Why do firms from developing countries bid higher? *Journal of International Business Studies, 42*(1), 128–151.

Huang, X. (2011). Corporate control by Chinese MNCs over their subsidiaries: Combining institutional and resources-based perspectives. *Labour and Industry, 22*(1/2 August/December), 169–192.

Huang, X. & Austin, I. (2011). *Chinese investment in Australia: Unique insights from the mining industry*. New York: Palgrave-Macmillan.

Hymer, S. H. (1976). *International operation of national firms: A study of direct foreign investment*. Cambridge: MIT Press.

Johanson, J. & Vahlne, J.-E. (2009). The Uppsala internationalization process model revisited: From liability of foreignness to liability of outsidership. *Journal of International Business Studies, 40*(9), 1411–1431.

Johanson, J. & Wiedersheim-Paul, F. (1975). The internationalization of the firm – Four Swedish cases. *Journal of Management Studies, 12*(3), 305–320.

Johnson, G., Scholes, K., & Whittington, R. (2008). *Exploring corporate strategy* (8th ed.). Sydney: Prentice Hall.

Johnson, G., Whittington, R., & Scholes, K. (2011). *Exploring strategy* (9th ed.). Sydney: Prentice Hall.

Kang, J.-K. & Kim, J.-M. (2010). Do foreign investors exhibit a corporate governance disadvantage? An information asymmetry perspective. *Journal of International Business Studies, 41*(8), 1415–1438.

Kayali, M. M. (2006). Real options as a tool for making strategic investment decisions. *Journal of American Academy of Business, Cambridge, 8*(1), 282–286.

Khanna, T. & Palepu, K. G. (2006). Emerging giants. *Harvard Business Review, 84*(10), 60–69.

Kogut, B. & Chang, S. J. (1996). Platform investments and volatile exchange rates: Direct investment in the U.S. by Japanese electronic companies. *The Review of Economics and Statistics, 78*(2), 221–231.

Kogut, B. & Kulatilaka, N. (1994). Options thinking and platform investments: Investing in opportunity. *California Management Review, 36*(2), 52–71.

Kojima, K. (1978). *Direct foreign investment: A Japanese model of multinational business operations*. London: Routledge.

Kojima, K. (1982). Macroeconomic versus international business approach to direct foreign investment. *Hitotsubashi Journal of Economics, 23*(1), 1–19.

Krychowski, C. & Quélin, B. V. (2010). Real options and strategic investment decisions: Can they be of use to scholars? *Academy of Management Perspectives, 24*(2), 65–78.

Li, P. P. (2003). Toward a geocentric theory of multinational evolution: The implications from the Asian MNEs as latecomers. *Asia Pacific Journal of Management, 20*(2), 217–242.

Liu, Q. (2006). Corporate governance in China: Current practices, economic effects and institutional determinants. *CESifo Economic Studies, 52*(2), 415–453.

Liu, Y. & Waldemar, A. P. (2011). China's emerging giants. Retrieved 1 February 2013, from http://www.businessforum-china.com/article_detail.html?articleid=7.

Liu, Y. & Woywode, M. (2013). Light-touch integration of Chinese cross-border M&A: The in?uences of culture and absorptive capacity. *Thunderbird International Business Review, 55*(4), 469–483.

Luo, Y. (1999). Time-based experience and international expansion: The case of an emerging economy. *The Journal of Management Studies, 36*(4), 505–534.

Luo, Y. & Tung, R. L. (2007). International expansion of emerging market enterprises: A springboard perspective. *Journal of International Business Studies, 38*(4), 481–498.

Luo, Y., Xue, Q., & Han, B. (2010). How emerging market governments promote outward FDI: Experience from China. *Journal of World Business, 45*(1), 68–79.

Mathews, J. A. (2006). Dragon multinationals: New players in 21st century globalization. *Asia Pacific Journal of Management, 23*, 5–27.

Meyer, K. E. (2004). Perspectives on multinational enterprises in emerging economies. *Journal of International Business Studies, 35*, 259–276.

Meyer, K. E., Estrin, S., Bhaumik, S. K., & Peng, M. W. (2009). Institutions, resources, and entry strategies in emerging economies. *Strategic Management Journal, 30*, 61–80.

Minerals Council of Australia. (2012). Life cycle of a mine. Retrieved 1 January 2013, from http://www.minerals.org.au/resources/gold/life_cycle_of_a_mine.

Moon, H.-C. & Roehl, T. W. (1993). An imbalance theory of foreign direct investment. *Multinational Business Review, 1*(1), 56–65.

Moon, H.-C. & Roehl, T. W. (2001). Unconventional foreign direct investment and the imbalance theory. *International Business Review, 10*, 197–215.

Morck, R., Yeung, B., & Zhao, M. (2008). Perspectives on China's outward foreign direct investment. *Journal of International Business Studies, 39*(3), 337–350.

OECD. (2008). OECD benchmark definition of foreign direct investment. 4th edition. Retrieved 1 February 2012, from http://www.oecd.org/dataoecd/26/50/40193734.pdf.

Ozawa, T. (1992). Foreign direct investment and economic development. *Transnational Corporations, 1*(1), 27–54.

Pan, Y. & Chi, P. S. K. (1999). Financial performance and survival of multinational corporations in China. *Strategic Management Journal, 20*(4), 359–359.

Peng, M. W., Wang, D. Y., & Jiang, Y. (2008). An institution-based view of international business strategy: A focus on emerging economies. *Journal of International Business Studies, 39*, 920–936.

Prahalad, C. K. & Hamel, G. (1990). The core competence of the corporation. *Harvard Business Review* (May–June), 79–91.

Rivoli, P. & Salorio, E. (1996). Foreign direct investment and investment under uncertainty. *Journal of International Business Studies, 27*(2), 335–357.

Rugman, A. M. (1986). New theories of the multinational enterprise: An assessment of internalization theory. *Bulletin of Economic Research, 38*(2), 101–118.

Rui, H. & Yip, G. (2008). Foreign acquisitions by Chinese firms: A strategic intent perspective. *Journal of World Business, 43*(2), 213–226.

Slangen, A. H. L. & Hennart, J.-F. (2008). Do multinationals really prefer to enter culturally distant countries through greenfields rather than through acquisitions? The role of parent experience and subsidiary autonomy. *Journal of International Business Studies, 39*(3), 472–490.

Sun, S. L., Zhang, Y., & Chen, Z. (2013). The challenges of Chinese outward investment in developed countries: The case of CITIC Pacific's Sino Iron Project in Australia. *Thunderbird International Business Review, 55*(3), 313–322.

Tam, O. K. & Yu, P. (2011). China's corporate governance development. In C. Mallin (Ed.), *Handbook on international corporate governance* (pp. 221–246). Cheltenham, UK and Northampton, MA: Edward Elgar.

Teagarden, M. & Cai, D. (2009). Learning from dragons who are learning from us: Developmental lessons from China's global companies. *Organizational Dynamics, 38*(1), 73–81.

UNCTAD. (2011). World investment report. Retrieved 1 February 2012, from http://www.unctad-docs.org/files/UNCTAD-WIR2011-Full-en.pdf.

Vahlne, J.-E. & Johanson, J. (2013). The Uppsala model on evolution of the multinational business enterprise: From internalization to coordination of networks. *Multinational Business Enterprise, 30*(3), 189–210.

Vernon, R. (1966). International investment and international trade in the product cycle. *Quarterly Journal of Economics, 80*(2), 190–207.

Weber, Y. & Tarba, S. Y. (2013). Integration in mergers and acquisitions: New perspectives. *Thunderbird International Business Review, 55*(4), 327–331.

Williams, C. & Martinez, C. A. (2012). Government effectiveness, the global financial crisis, and multinational enterprise internationalization. *Journal of International Marketing, 20*(3), 65–78.

Williamson, O. E. (1981). The economics of organization: The transaction cost approach. *The American Journal of Sociology, 87*(3), 548–577.

Williamson, O. E. (1991). Strategizing, economizing, and economic organization. *Strategic Management Journal, 12*(Special issue), 75–94.

Yiu, D. W., Lau, C., & Bruton, G. D. (2007). International venturing by emerging economy firms: The effects of firm capabilities, home country networks, and corporate entrepreneurship. *Journal of International Business Studies, 38*(4), 519–540.

Zaheer, S. (1995). Overcoming the liability of foreignness. *Academy of Management Journal, 38*(2), 341–363.

Zhang, G. (2009). On the development and evolution of Chinese fundamental policies on outward direct investment. Retrieved 7 September 2009, from http://www.caitec.org.cn/c/cn/news/2009-08/10/news_1554.html.

Zhang, Y., Zhang, Z., & Liu, Z. (2007). Choice of entry modes in sequential FDI in an emerging economy. *Management Decision, 45*(4), 749–772.

Zollo, M. & Meier, D. (2008). What is M&A performance? *Academy of Management Perspectives, 22*(3), 55–77.

3

How Does Liability of Foreignness Impact the Behaviour of Chinese MNCs? A Case Study of Sino Iron Project

Introduction

On 11 November 2012, CITIC Pacific (CP), a Chinese-controlled and publicly listed Hong Kong-based company, announced that its wholly owned Australian subsidiary, CITIC Pacific Mining (CPM), had load-commissioned its first production line out of a total six in its Sino Iron Project (SIP) in Western Australia. Although this can be regarded as a milestone for the SIP, the project had been delayed by nearly four years, as CP had initially expected start-up in early 2009. The costs of the SIP, partially due to this four-year delay, had increased from the initially projected US$2.5 billion to US$7.8 billion by the end of June 2012. The general media is expecting the final costs of the SIP to reach US$10 billion. The delay in production has been largely attributed to its long and costly legal battle with its Australian partner, Mineralogy, over several major issues, such as mining royalty, third-party access to the CPM-built port facilities and environmental protection (Chenoweth, 2014).

What contributed to this nearly four-year project delay and potentially 300 per cent cost blowout for a major mining project in Australia? This question has been asked by many academics, business executives and government officials in Australia, and particularly in China. This chapter attempts to address the question from the perspective of "liability of foreignness" (LOF) (Zaheer, 1995) using a pluralistic theoretical approach.

Such a study is warranted for both theoretical and practical reasons. Theoretically, Chinese OFDI is relatively new phenomenon despite the fact that China became the third largest FDI home country in the world in 2013 (Chenoweth, 2014; UNCTAD, 2014). Although the concept of LOF has been long developed (Zaheer, 1995) to explain difficulties in conducting international business, most studies have used this concept to explain the behaviour of MNCs from developed countries doing business in developing or developed countries (Ministry of Commerce, 2014).

Chinese investors, particularly state-owned enterprises (SOEs), may incur liabilities and deploy mitigating strategies different from those of developed countries due to their state ownership, lack of international experience and national identity. It is not clear how these factors may affect the extent of LOF, thus impacting FDI decisions and management of Chinese subsidiaries in foreign countries. From a practical perspective, the findings of the present study can help Chinese business executives understand the factors contributing to their LOF and influencing their FDI performance, thereby improving the quality of their FDI decisions and management. For business executives in host countries, the findings can assist them to understand LOF suffered by Chinese investors and thus to be better prepared in competing or cooperating with Chinese investors, either as subcontractors, joint venture (JV) partners, or as consultants.

Reviewing literature on liability of foreignness

MNCs competing in foreign countries incur additional costs which are not or are only partially applicable to local firms (Luo et al., 2002; Zaheer, 1995). Conceptually, LOF is defined as "all additional costs a firm operating in a market overseas incurs that a local firm would not incur" (Zaheer, 1995, p. 343). More specifically, Mezias (2002, p. 266) refer to LOF as "phenomena that cause foreign firms to incur costs domestic firms do not, incur costs to a greater extent than domestic firms do, or be denied benefits only domestic firms are eligible to receive".

The original idea of LOF can be traced back to Hymer (1976) who argued that MNCs need to have monopolistic advantage to overcome their disadvantages in foreign countries in order to compete successfully. Based on these definitions, LOF has two important characteristics. First, it is *dyadic* as it involves both the home and host country (Australian Bureau of Statistics, 2013), and second, it is *relative* due to its essential reference to indigenous firms (Huang & Austin, 2011). Eden and Miller (2001) used the metaphor of "stranger in a strange land" (p. c4) for describing these two characteristics. This also implies that the degree of LOF a MNC suffers varies with the specific context, such as firm, industry, home country and host country (Zaheer, 1995).

What are the sources of LOF? Zaheer (1995) pointed out four broad sources of LOF: (1) costs related to spatial distance (e.g. travel, transportation and coordination over distance and across time zones); (2) firm-specific costs stemming from the firm's unfamiliarity with the environment and lack of roots in the host country; (3) costs arising from the host country environment, such as the lack of legitimacy of foreign firms, administrative barriers and economic nationalism; and (4) costs caused by regulations in the home country, such as the restrictions on high-technology sales to certain countries imposed on some Western multinationals. Eden and Miller (2004) excluded relative production cost from LOF and grouped the sources of LOF into three categories: *unfamiliarity hazard, relational hazard* and *discrimination hazard*. Therefore, LOF is the combined consequence of institutional distances (regulative, normative and cognitive) (Scott, 1995), MNCs' international experience and network, as well as the effect of the external environment of both host and home countries on their international operations. It is both resources-based and institution and society-embedded.

Several theories have been used in extant literature to explain why MNCs suffer LOF, including Resource Based View (RBV), institutional theory, learning theory, social networking and international expansion (Ministry of Commerce, 2014). However, these theories focus on different aspects of LOF. While RBV and institutional theory attempt to explain the roots of LOF, organisational learning and social networking theories concentrate on how to mitigate LOF, and international expansion theory focuses on the context and dynamics of LOF.

Another central issue in the study of LOF is to identify effective mechanisms, or strategies, to mitigate LOF. To do so, foreign firms can use two broad approaches (Zaheer, 1995): bringing in their firm-specific advantages, such as technologies, brand names or managerial know-how, to their foreign subsidiaries, or mimic the successful practice of local firms in the host country. They can be defensive or offensive (Luo et al., 2002). The former refers to strategies that reduce the cost of hazards from the host country, including contract protection, parental control, parental service and output standardisation, while the latter refers to strategies that improve relations between the host country and the firm, covering local networking, resource commitment, legitimacy improvement and input localisation.

Other LOF-mitigating mechanisms include entry mode selection (export, JV or acquisition) (Klinger, 2013), first mover (Eden & Molot, 2002) and boundary spanning (Elango, 2009). For example, Eden and Miller (2001) argued that wholly owned subsidiaries should be used as entry mode for countries, where institutional distance and ethnocentrism are low, by MNCs with good reputation, high levels of legitimacy and prior experience in international operations.

A number of articles have focused on the LOF suffered by MNCs from emerging markets. For example, Miller et al. (2008) argued that MNCs from emerging markets entering developed countries have to "compete from a position of double disadvantage: The firms incur additional costs of doing business abroad, and in addition they are often resource poor compared to domestic firms in developed market" (p. 646). However, MNCs from emerging countries can draw on their ethnic identity and cooperation with ethnically similar firms to compete in the host developed country. Klossek, Linke and Nippa (2012) studied how Chinese subsidiaries in Germany mitigated their LOF and found that they employed several mechanisms, including due diligence, sharing responsibilities, reputation building and prior experience in international operations. However, their study did not elaborate on how Chinese MNCs employed such mechanisms to mitigate their LOF in Germany.

Research method

Case selection

This chapter aims to explore the LOF suffered by Chinese MNCs, particularly SOEs in the Australian mining industry. As Chinese investment is a contemporary phenomenon, a case study is appropriate for such an investigation (Yin, 2003). The Sino Iron Project (SIP) was selected for several reasons:

(1) It is by far the largest Chinese Greenfield investment in the Australian mining industry. It was also the first international venture by CP and the first international mining project by the CITIC Group after it became the majority shareholder of CP in December 2008. The project was established in 2006 and was one of the earliest major Chinese Greenfield investments in the Australia. As the first production line was load-commissioned in November 2012, the degree of LOF and effect of CP's strategies and actions to mitigate its

LOF during its entire developmental stage have already played out and thus can be studied in detail.

(2) Using Greenfield to enter the international market incurs a higher level of LOF than acquisitions. Coupled with the large scale and complexity of the project, the SIP is an excellent case study of LOF as it can provide a comprehensive coverage of the sources of LOF and their effect on Greenfield investment in the Australian mining industry.

(3) By the end of 2013, Australia had become the fifth largest destination for Chinese OFDI, and more than 60 per cent of Chinese investment in Australia was in the mining industry (Ministry of Commerce, 2013, 2014). Thus, this study can help other Chinese investors understand the LOF in investing in the Australian mining industry and develop their strategies for mitigating these liabilities.

Data collection

As part of a large research project on Chinese OFDI in Australia, this chapter focuses on the causes of the SIP's poor performance from the perspective of LOF. We conducted five interviews with five senior and middle managers, and we undertook a content analysis using the publicly available information from three sources: (1) *Factiva*, an e-database containing major Australian newspapers, including *The Australian*, *The Australian Financial Review* and *Western Australia Business News* since the start of 2006 when CP started investing in Australia. A total of 168 newspaper reports were collected; (2) CP's company reports and announcements on its company website (www.citicpacific.com) since 2006; and (3) *Baidu*, the most popular Chinese Internet search engine for online Chinese media reports. In addition, we also used Internet search for collecting supplementary information where necessary.

Case description

CPM was established in 2006 and is a wholly owned subsidiary of CP, with the CITIC Group holding 28.78 per cent of its share before 2009.

The purpose for setting up CPM was to manage the SIP. In April 2006, CP acquired the right to mine 1 billion tons of iron ore (magnetite) at Cape Preston, which is located 100 kilometres south of Karratha (1,535 kilometres north of Perth, Western Australia). The SIP was designed to be the biggest magnetite ore producer in the world with an annual production capacity of 24 million tons of iron ore concentrate. The originally expected cost of the project was estimated to be US$2.5 billion with the first production line in operation in early 2009 (CITIC Pacific, 2006).

The project consists of mining operations, an ore beneficiary plant (crushing, grinding and concentrating), a dewatering plant, a 30-kilometre slurry pipeline, a 450-megawatt gas-fired power plant, a 51 million kilolitre desalination plant, port and transshipment facilities and other infrastructure. This was the first international project for CP and the first mining project for the CITIC Group.

However, the project has suffered a severe cost blowout, with estimated costs close to US$10 billion. The timeline of the major events of the project is outlined in Table 3.1.

Table 3.1 The timeline of the Sino Iron Project

Date	Event
31 March 2006	CP invested US$215 million to acquire rights to mine 1 billion tons of iron ore (magnetite).
June 2006	Barry Fitzgerald was appointed as CEO and Chief Engineer of CPM.
24 January 2007	CPM entered into the contract with MCC for US$1.106 billion, announced the total costs of the SIP was expected to be US$2.5 billion with the first production line completed in June 2009.
20 August 2007	CP entered into the Supplemental Agreements with MCC in relation to the adjustment to the scope of the works as set out in the Original Contract and the revision of the Contract Sum to US$1.75 billion and sold 20% interest of SIP to MCC for US$448 million.
November 2007	CP acquired further rights to mine 1 billion tons of iron ore for US$200 million.
17 April 2008	CP announced more than US$1 billion blowout in costs to US$4.2 billion.
24 December 2008	CITIC Group paid HK$11.625 billion (US$1.5 billion) to CP to become the majority shareholder of CP, increasing its shareholding from 28.78% to 57.56%.
6 January 2009	CPM signed an agreement for seven years with Apache and Santos for supplying LNG to CPM's power plant at the highest gas price for the region.
8 April 2009	Mr Larry Yung Chi Kin resigned as Director and Chairman of the Company due to the foreign currency exchange hedging. So did the Managing Director Mr Henry Fan Hung Ling.

Table 3.1 (Continued)

Date	Event
10 May 2009	Dr Dongyi Hua was appointed as the Chairman of CPM with other four senior executives from CITIC Construction and arrived in Australia in September 2009.
4 May 2010	CP signed a supplemental agreement under which it agreed to pay an additional amount of US$835 million to MCC and announced cost blow out to US$5.2 billion.
21 May 2010	Barry Fitzgerald resigned as the Managing Director and Chief Engineer of CPM.
18 August 2010	CP stated it expected the first production line should be ready for operational testing by the end of 2010, and the first concentrate shipment in the first half of 2011.
19 January 2011	CP updated to the market that all components needed for the operation of the first production line should be ready for commissioning as an integrated system by the end of July 2011, and export of iron ore was expected to be in the latter part of 2011.
18 July 2011	CP announced that it expected the production and export of iron ore concentrate in the first half of 2012.
30 December 2011	CP paid an additional US$822.1 million to MCC for the completion of the first two production lines and the common facilities, increasing the contract sum from US$2.585 billion to US$3.407 billion. It also announced that the first production line was expected to be completed by the end of August 2012. The project costs were expected to be more than US$6.0 billion.
13 April 2012	Acquiring further right to mine 1 billion tons of iron ore, totalling 3 billion tons of iron ore (magnetite). The project expenditure reached US$7.1 billion by the end of 2011.
19 November 2012	CP announced that the load commissioning of the first production line and the common facilities of the SIP had been achieved and the concentrator plant had begun to produce iron ore concentrate.
2 December 2013	First magnetite concentrate was shipped out from the port.
12 July 2014	First 1 million tons of magnetite concentrate was shipped out and the second phase of construction started.

Findings and discussion

Many factors have contributed to the massive cost blowout and dramatic project delay of the SIP. In this study, we primarily aim to identify the LOF suffered by CP using Eden and Miller's (2004) three types of sources of LOF and describe key mitigating strategies used by CP.

Poor preparation for entering the Australian mining industry

CP's entry into the Australian mining industry was not well planned. Clive Palmer, who bought the right to mine the magnate resources at Cape Preston in 1986, had been attempting to sell the mining rights, but failed to find a buyer before 2006. The main reason was that magnetite is lower in iron content and higher in production costs compared to hematite (another type of iron ore, usually called "direct shipping ore") as it requires processing into concentrate before it can be economically used as raw material by steel mills.

The time spent on project selection by CP was very short. The project was first introduced to another Chinese SOE early last decade. A former Vice President of this Chinese SOE then introduced it to CP in late 2005 after he retired from the SOE. The agreement to buy mining rights for 1 billion tons was signed at the end of March 2006, only several months after this project was introduced to CP, and "coincidently" one day before former Chinese Premier Wen Jiabao visited Australia. Given the complexity and scale of this project, and more importantly, the fact that CP had little experience in mining operations, in particular, in Australia, it seems that CP rushed into the project without appropriate preparation and analysis, possibly due to political pressure from the Chinese government to sign the agreement before its Premier visited Australia.

The entry mode used by CP was also inappropriate judging from its low level of experience in international operations and in the mining industry. CP selected Greenfield as the entry strategy into the Australian mining industry, particularly into magnetite mining. The Greenfield mode incurs a higher level of LOF than acquisitions or joint ventures as it involves greater uncertainty and a higher level of risk (Klinger, 2013). In other words, the decision to use Greenfield to enter magnetite was very risky. The unfamiliarity hazard in this case contributed to the selection of this particular project, using Greenfield as entry mode.

Chinese standards in project design and costing

The LOF contributed by the unfamiliarity hazard is also reflected in the project design and feasibility study. Project design and costing are widely regarded as important steps for investing in a mining project. Pre-feasibility and feasibility studies are two crucial studies to help decide if the project is economically viable. While the quality of the pre-feasibility and feasibility study reports in this particular case are not available for inspection, the time spent on them was less than six months based on the information provided in its announcements to the Hong Kong Stock Exchange; a much shorter period compared with another Australian miner who took more than one year to complete a feasibility study in a similar, but less complex, magnetite project. The mining agreement was approved by the Australian Foreign Investment Review Board (FIRB) in June 2006. CP had already selected its lead contractor and started its construction as reported in its 2006 Annual Report:

> We have signed a general construction contract with China Metallurgical Group Corporation ("MCC") (in January 2007). Work on the development of mining, beneficiation plant, pellet plant, desalination plant, power station and material handling system has begun. First shipment of product is expected to be in 2009.
>
> (CITIC Pacific, 2006, p. 5)

The pre-feasibility and feasibility were conducted by a Chinese-led consortium. The costing and standards were primarily based on those in China. As expressed by a veteran Chinese manager, the feasibility study in China is "often very simple, with the purpose for getting approvals from the superior organization or government". Consequently, many designs and specifications had to be substantially modified to conform to Australian standards in the construction stage, which caused project delays and increased costs (CITIC Pacific, 2010).

Difficulties and costs in managing major contractors

The LOF caused by *relational hazard* for the project is vividly reflected in the contract management. Used properly, contracts are an effective approach to mitigating LOF (Sun et al., 2013) as contract protection can reduce the complexity and bureaucratic costs of inter-firm coordination (Al Najjar, 1995) and alleviate conflict derived from heterogeneous goals, culture and strategies (Geringer & Hebert, 1989). However, CP encountered many issues in dealing with its two major contractors,

Metallurgical Corporation of China (MCC) – a large Chinese SOE – as the lead contractor for construction, and Mineralogy, an Australian company, as the port management agency and its mining royalty owner.

Dealing with the project's lead contractor, MCC, became a major problem and incurred much extra cost for CPM. In January 2007, MCC was selected as the lead contractor at a price of US$1.1 billion, primarily based on the company's low cost bid. MCC is responsible for the project's infrastructure, including "the design, construction and installation of the primary crushing plant, concentrator, pellet plant, material handling system, camp and other auxiliary infrastructure facilities. It also includes procurement of certain mining equipment" (CITIC Pacific, 2006, p. 14). Although MCC was regarded as having "extensive experience and expertise" in the design and construction of similar minerals processing plants in China and some developing countries, such as Brazil and Iran (CITIC Pacific, 2006), it had little experience in undertaking a large-scale construction project in developed countries such as Australia. The MCC contract price was increased three times from the original US$1.1 billion in 2006 to US$1.75 billion in 2007, US$2.58 billion in May 2010 and US$3.407 billion at the end of 2011; even the design and construction of the pellet plant from the original MCC contact was shelved. MCC has been blamed for the increases in the contract price due to its unfamiliarity with the relevant Australian policies (CITIC Pacific, 2011, 2012). "MCC told us [CP] that they underestimated the complexity and the amount of work involved in constructing and commissioning a project in Australia" (CITIC Pacific, 2011, p. 5). These policies include immigration policy, environmental protection and occupational health and safety in the workplace.

Many factors have contributed to more than 300 per cent cost blowout for MCC. Three major factors are (1) the company based its costing on cheap Chinese skilled labour; (2) it took for granted that the Australian workforce is as flexible as the Chinese in working hours – a major method used in China to beat the project deadline; and (3) it underestimated the cost for environmental protection.

Labour costs were very high in Australia and have rapidly increased since 2006 due to the demand from mining and construction. According to the Australian Bureau of Statistics, the weekly wage in Western Australia increased 41 per cent between 2006 and 2011 (Australian Bureau of Statistics, 2013). Moreover, the average hourly cash earnings for those working in the mining industry was US$52.30, more than double of those working in the retail trade (US$25.20) in May 2012 when the national minimum wage was US$15.51 per hour.

Considering other welfare factors, such as annual leave, sick leave and holidays, it is estimated that the Australian labour wage/salary in Western Australia could be 17 times that of their Chinese counterparts (Audeyev et al., 2014), while the accommodation cost for labourers was about nine time higher than that in China. Labour is a major part of costs in the construction industry. MCC's calculations were based on the Chinese wage standard. With special permission from Australia's Department of Immigration, it was allowed to bring 400 Chinese skilled labourers, rather than over 3,000 in its original design. Moreover, strict industry regulations had made the flexible use of workforce uneconomical. Both have contributed substantially to the project delays and cost blowouts (Sun et al., 2013). The high standard of Australian environmental protection also contributed substantially to the cost increase. For example, building a bridge in China would cost about RMB5 million in China, but it cost A$50 million in Australia primarily due to marine protection (Chen, 2010a).

Consequently, CPM made considerable efforts in dealing with MCC, including the revision of the contract and signing a new penalty agreement at the end of 2011 for further project delays after 31 August 2012 for the load commission of the first production line. The massive extra cost in dealing with MCC demonstrated how a new context can contribute to the LOF. Although CP knew MCC well before 2006 when the contract was signed, dealing with MCC in the new international context was a loss-loss situation for both parties involved. It was like putting "new wine into old bottles" – the bottles break and the wine runs out!

CP has also suffered substantially in its relationship with its mining right owner and port management agency, Mineralogy. CP bought the mining rights from Mineralogy in 2006. As part of the mining right acquisition, CP was required to pay Mineralogy royalties which are calculated in two parts. Part A is a fixed A$0.35 for every ton of ore mined. Part B is calculated based on the percentage of the then Australian benchmark iron ore price (Mount Newman benchmark price). However, the long project delay caused the deterioration of CP's relationship with Mineralogy. The abolishment of the Australian iron ore benchmark price has also added to the further deterioration of this relationship. A long legal battle between CP and Mineralogy began in 2010 and escalated in mid-2014 as Mineralogy issued a default note to CP, threatening to shut down its port operations in August 2014 (Chenoweth, 2014).

Other major components of the project, except mining which is under the direct management of CPM, were contracted out to several Australian and international firms, such as IDE Technologies for the desalination plant, AE&E for the power plant, JV between NRW Holding and VDM group for Port Breakwater. These contracts have protected the project costs as the expenditure was specified in the contracts. However, the mining component under the direct responsibility of CPM also exposed it to the rising costs of labour and mining equipment.

Moreover, CPM signed a gas supply contract with a local gas company for its power plant at the time of short gas supply caused by the mining boom. Under this seven-year gas supply agreement, CPM is committed to purchase 65 terajoules of gas a day and pay a minimum lump sum annually from late 2011 for a quota of gas consumption, regardless of whether it uses the gas or not (Chenoweth, 2014; Klinger, 2013). It is estimated that CPM lost tens of millions of dollars between 2010 and 2013 due to its project delay (Chenoweth, 2014).

Additional costs in corporate management

Internal organisation cost is another form of *relational hazard* (Eden & Miller, 2004). The LOF caused by this form of relational hazard has been mainly reflected in the areas of trust and organisational culture.

In order to foster trust, an Australian-led management team was recruited immediately after the acquisition of the project by CITIC Pacific in 2006. Barry Fitzgerald, a mining veteran who had more than 30 years' experience in the mining industry, was appointed as the CEO and Chief Engineer of CPM in 2006. This local appointment was to "mimic" the successful practice of local firms (Zaheer, 1995) as an effective way of reducing LOF caused by CP's unfamiliarity with the mining operations in Australia and its relational hazard with local business and governments. As mentioned above, all major components (except mining operations) of the project were contracted out to Australian and international firms under Fitzgerald's leadership. Nevertheless, early signs of project delays and cost pressures emerged as early as 2007, undermining trust in the project as CP became concerned about management's competence and conflict of interest in managing this project.

Shortly after the CITIC Group took control of CP in April 2009, five senior managers were sent from CITIC Construction to CPM. They held several key positions at CPM, including the Executive Chairman of the Board and Chief Operations Officer. The main purpose of this move was to transfer project management experience to CPM to speed up the project progress, in addition to addressing the lack of trust. This is another commonly used approach to mitigate LOF (Zaheer, 1995). These five senior managers had gained their project management experience in China and some developing countries, particularly in Africa.

Fitzgerald resigned and left CPM in May 2010, less than eight months after the arrival of five Chinese senior managers and two weeks after CPM announced that it agreed to pay an additional US$8.35 billion to MCC, the project budget being lifted to US$5.2 billion. The media widely speculated that he had been a scapegoat and had jumped before being pushed.

At the beginning of its appointment, the Chinese management team attempted to apply a Chinese approach in managing organisational culture and human resources. For example, the CITIC Group's slogan was displayed in both Chinese and English at the reception of CPM Head Office. Managers and employees were asked to learn about Chinese culture. However, this caused a strong resistance, particularly from the Australian middle managers, as one middle manager complained during the interview: "We had our own corporate culture before, but suddenly it's all gone. Why we are asked to learn about Chinese culture, and not the Chinese learn our culture?" Workforce turnover was high in 2010. This prompted the shift of management approach back to local practices as one senior manager commented: "I am back to square one. I would like to learn what they (Australians) are doing first, then implement the changes I want".

Having said this, CPM has employed local managers in several key positions in which Chinese managers lack experience and knowledge, such as HR, environmental protection, safety and occupational health, external affairs and legal services.

Discriminatory government regulations and administrative procedures

The third major source of LOF suggested by Eden and Miller (2004) is *discriminatory hazards*, caused by "the discriminatory treatment inflicted on the foreign firm relative to local firms in the host country".

Like any other FDI, CP's investment in Australia is generally welcomed and foreign firms are not discriminated by the Australian government at the federal and state levels. However, the perceived discrimination by a foreign firm can adversely affect its behaviour and performance. In the case of CP, it deliberately opted to lease all the land and the port on its mining and operations site from Mineralogy, rather than acquiring them as a way of avoiding potential problems with the Foreign

Investment Review Board in its approval of the investment proposal (Chenoweth, 2014). This has caused a substantial problem at the late stage of the project in coordinating its value chain in Australia, due to its legal battle with Mineralogy for control and the uncertainty in shipping out its iron ore to Asia.

There could be other discriminatory hazards from both relative and dyadic perspectives. From a relative perspective, for example, the Australian government permitted MCC to hire 400 foreign workers out of the 4,000 workers it required (10 per cent of the workforce). However, it approved Hancock Prospecting (an Australian mining company) in May 2012 to hire up to 1,715 foreign workers for their 8,000-worker project (about 20 per cent of the workforce) (Bowen, 2012). This might be regarded as a discriminatory hazard as the government allowed a local mining company to hire a greater percentage of foreign workers than MCC, giving Hancock Prospecting a competitive advantage in costs due to cheaper labour. From a dyadic perspective, some government regulations in Australia, for example, the English Proficiency Test for imported skilled labour, have also substantially contributed to the project delay and consequently cost blowout, particularly in the Sino Iron project (Metallurgical Corporation of China, 2014). This has become the biggest obstacle for Chinese MNCs, whereas it has not been a problem for English-speaking countries, such as India, South Africa and the Philippines. Moreover, some regulations are not explicitly codified or have socially embedded components, particularly in environmental protection, community engagement and industry regulations. Administratively, visa processing is another liability for Chinese MNCs. All these liabilities can be illustrated in the statement of CP's Chairman:

> [T]here are strict commissioning requirements, ranging from the certification of construction completion documents by *licensed Australian engineers* to meeting stringent safety regulation standards... commissioning of control systems... must be done by *qualified Australian electricians*.... commissioning also requires the assistance of equipment service providers, of which there are hundreds, and the on-going need to *process the visas of these workers* has continued to be complicated and time consuming.
>
> (CITIC Pacific, 2012, p. 8 emphasis added)

Discussion

CP's venturing into the Australian mining industry with Greenfield as entry mode has typified the LOF suffered by an inexperienced firm from an emerging country investing in a developed country. It is a typical example of "stranger in the strange land" (Klinger, 2013) in terms of CP's identity as seen by its key Australian stakeholders (Australian business communities, media, public and governments) as well as its unfamiliarity with the Australian business environment and contexts. Consequently, it has incurred a four-year project delay and in excess of 300 per cent cost blowout. Although delays and cost blowouts are not uncommon in developing major mining or oil and gas industries in Australia, they are much longer and more costly at CPM than other local firms. Besides several contextual factors that are out of the company's control, such as the rising costs of labour, equipment and materials, LOF has been a major factor contributing to the project delay and cost blowout.

Both RBV and the institutional perspective can be used to explain the high level of LOF suffered by CP. The company had no experience in mining before it invested in Australia; thus, it was unfamiliar with the country's mining operations and business environment. It started this project "knee deep in the Big Muddy" (Chen, 2010a, p. 646) and competed from "a position of double disadvantage" (p. 646). The poor preparation and feasibility study partly reflected CP's unfamiliarity caused by its lack of experience.

CP also incurred additional costs from a relational hazard standpoint as it was an outsider in the global business network of mining projects, including engineering, procurement and construction. According to the revised Uppsala model (Johanson & Vahlne, 2009), firms operating outside the local business networks will incur a "liability of outsidership" due to the fact that they are not privy to information from inside the business circle.

The institutional perspective also sheds much light on the LOF suffered by CPM. Although CP is listed on the Hong Kong stock exchange, it was still regarded as a Chinese firm with a strong connection with the Chinese government even before the CITIC Group took a majority share in 2009. The legitimacy of Chinese investors, subjected to the "country of origin effects" (Vernon, 1977) in the Australian mining industry, has been challenged by the Australian media and the public (Huang & Austin, 2011). In 2009, there was a strong negative sentiment among the Australian media and the public to Chinese investment (Huang & Austin, 2011), partly due to their lack of understanding of China and Chinese firms, the cognitive pillar of institution. From a regulative pillar view, the Australian government was reluctant to deviate from its administrative procedures and policies to CPM's requests, particularly with regard to importing Chinese skilled labour to speed up the project, although there was a considerable labour shortage in Australia.

To mitigate its LOF, CPM had primarily used defensive strategies (Sun et al., 2013), including contract protection and the use of parental expertise to reduce its vulnerabilities in Australia; in particular, it guarded against conflicts of interest when dealing with foreign entities and unexpected changes in Australia. This is appropriate according to the Importance-Liability-Matrix developed by Sofka and Zimmermann (2008), as the Australian market was of low importance to CPM and the company had a great need to protect its multi-billion dollar project from cost increases. There is evidence that CPM effectively used contract protection to protect itself against unexpected changes in construction, such as AE&E failing to build the power plant on schedule. However, major problems in adopting its strategies included dealing with the lead contractor, MCC, and its partner, Mineralogy.

The home country institutions have also substantially influenced how CP responded to its LOF in this project. The intuitional forces included political pressure, trust and taken-for-granted assumptions. First, the acquisition agreement was signed 1 day before the former Chinese Premier Wen Jiabao visited Australia. From the perspective of Chinese culture, this was regarded as a sign of good faith placed by the Chinese government in Australia, and a fruitful outcome of the official visit by the Chinese Premier to Australia from a political perspective. This may have reduced the time required for CP to deliberate and select this project. Trust was a key reason behind the appointment of the Chinese senior management team to CPM in 2009. The case also shows that the trust between the parent company and senior foreign managers at its foreign subsidiary are more likely to become problematic as the subsidiary's performance deteriorates.

Assumptions also played a part in how CPM and MCC behaved to reduce their LOF. Chinese MNCs have relied on the Chinese government in many occasions of OFDI, particularly in developing countries, to gain favourable treatment, such as importing labour, customs clearances and visa issuing. However, the Australian government was very reluctant to bend its administrative procedures and policies in response to CPM's requests. "In Australia, no one listens to our requests; or they [the Australian government officials] politely listen to us, but in fact strictly follow the procedures" commented the Executive Chairman of CPM (Chen, 2010b). Therefore, much effort had been spent on meeting with the Australian government officials, but with little effect on gaining favourable treatment from the Australian governments.

The massive cost blowout and dramatic project delay have also raised a question regarding corporate governance at CP. Besides the loss of HK$15.5 billion (about US$2.0 billion) in its foreign currency hedging in 2008, the selection of the project and Greenfield entry mode have reflected a weakness in corporate decision-making. The former chairman of CITIC Pacific was Larry Yung, a son of Rong Yiren, the former Vice President of China. Larry was the key decision-maker for the currency hedging in 2008. The upward drifting target of project costs and production schedule, particularly over the past three years, is another example of weak corporate governance in terms of transparency and information disclosure. The interaction between poor corporate governance and unfamiliarity has worsened the project outcome.

Conclusion and implications

This chapter explored the LOF suffered by Chinese MNCs in the Australian mining industry. Using the Sino Iron Project as a case study, we attempted to identify the specific LOF suffered by CPM in developing the project to date and the main strategies used by CPM in mitigating its LOF; we attempted to explain why such additional costs were incurred by the project and certain mitigating strategies were used by CPM. Several conclusions can be drawn based on the findings of this chapter.

Unfamiliarity with the Australian business environment and contexts, and particularly the local mining industry, has been a major source of LOF suffered by CPM. Consequently, CP made uninformed decisions in its project selection and entry mode. The root of this unfamiliarity can be attributed to the lack of international experience, both in Australia and in the mining industry, by CP and its leading contractor, MCC. Coupled with this unfamiliarity, the weak corporate governance at CP was another important cause for the failure of the SIP.

CP also suffered LOF due to the relational hazards generated from its inter-firm and intra-firm relationships. From the inter-firm perspective, both CP and MCC were viewed by most firms operating in the Australian mining industry as outsiders. Although both the CITIC Group, the majority shareholder of CP, and MCC are Chinese SOEs, the cost involved in dealing with MCC contributed substantially to the cost blowout. This was CP and MCC's first project together; there was no previous relationship between the two companies, thus they lacked the understanding and trust that could be expected between the parties. This relational cost was generated by the new international context. The long legal battle between CP and Mineralogy is another relational hazard that CP has suffered as part of its LOF in Australia. From the intra-firm perspective, trust in the management team and cultural clashes led to additional costs.

The SIP has also experienced discriminatory hazards. This is clearly seen in MCC's attempt to recruit Chinese workers for its project. The permit given to the local mining firm allowed it to hire 20 per cent of its workforce from overseas, whilst MCC was only permitted to import about 10 per cent. From a dyadic perspective, the English Proficiency Test for imported workers is another example in this regard. The fear of being discriminated against by the Australian government for its investment caused CP to behave irrationally in its acquisition decision as evidenced by its deals with Mineralogy for the land ownership at its operational site and port.

CP has mimicked local practices and engaged corporate competences in mitigating its LOF suffered in the project. Nevertheless, the use of such mitigating mechanisms was influenced by the home country's institutions, culture and by the organisational resources of the parent company.

Several implications can be suggested based on the findings of this chapter for both Chinese managers and business executives in host countries. For Chinese business managers, a solid preparation and analysis before making investment decision is indispensable. From a LOF perspective, and considering that LOF operates at a dyadic and relative level, these preparations and analyses should examine relevant factors in both home and host countries at three levels: national, industry, target organisation and/or project. Investments in the mining industry, moreover, need to adopt a process view of the mining cycle (exploration, construction, production, de-commissioning) in order to fully examine the economic benefits of an investment in the project life cycle. Such preparations and analyses can help make decisions on the investment, entry mode and strategies for mitigating LOF.

Business executives in host countries should understand the LOF suffered by Chinese MNCs and its sources in order to better prepare them to cooperate with these Chinese MNCs in their foreign operations, or to work within a Chinese MNC in terms of assisting Chinese MNCs to lower their LOF, thus improving organisational performance.

References

Al Najjar, N. I. (1995). Incomplete contract and the governance of complex contractual relationships. *AEA Papers and Proceedings, 85*(2), 432–436.

Audeyev, S., Carroll, M., & Allman, D. (2014). Corporate social responsibilities. In Z. Ma (Ed.), *How to invest in Australia* (pp. 192–203). Beijing: China Commerce and Trade Press.

Australian Bureau of Statistics. (2013). Australian Social Trends. Retrieved 1 December 2014, from http://www.abs.gov.au/AUSSTATS/abs@.nsf/Lookup/4102.0Main+Features10April+2013.

Bowen, C. (2012). First enterprise migration agreement approved. Retrieved 1 December 2013, from http://www.minister.immi.gov.au/media/cb/2012/cb187050.htm.

Chen, Z. (19 April 2010a). The beginning of love experience between CITIC Pacific and Australian iron ore. *Caixin.* http://magazine.caixin.com/2010-04-18/100136268.html.

Chen, Z. (2010b). Initial love experience with Australian iron ore by CITIC. Retrieved 1 December 2011, from http://magazine.caixin.com/2010-04-18/100136268_all.html#page2.

Chenoweth, N. (24 July 2014). The real deal with Clive. *The Australian Financial Review,* pp. 56–57.

CITIC Pacific. (2006). Annual report. Retrieved 1 November 2012, from http://www.citicpacific.com/report/annual/2006/citic06ar_e.pdf.

CITIC Pacific. (2010). Annual report. Retrieved 1 November 2012, from http://www.citicpacific.com/report/annual/2010/citic10ar_e.pdf.

CITIC Pacific. (2011). Annual report. Retrieved 1 November 2012, from http://www.citicpacific.com/report/annual/2011/citic11ar_e.pdf.

CITIC Pacific. (2012). Annual report. Retrieved 1 November 2012, from http://www.citicpacific.com/report/halfyear/2012/Interim_2012_Full_Version_e.pdf.

Eden, L. & Miller, S. (2001). Opening the black box: Multinationals and the costs of doing business abroad. *Proceedings of the Academy of Management,* C1–C6.

Eden, L. & Miller, S. (2004). Distance matters: Liability of foreignness, institutional distance and ownership strategy. *Advances in International Management, 16,* 187–211.

Eden, L. & Molot, M. A. (2002). Insiders, outsiders and host country bargains. *Journal of International Management, 8*(4), 359–388.

Elango, B. (2009). Minimizing effects of "liability of foreignness": Response strategies of foreign firms in the United States. *Journal of World Business, 44*(1), 51–62.

Geringer, J. M. & Hebert, L. (1989). Control and performance of international joint ventures. *Journal of International Business Studies, 20*(2), 235–254.

Huang, X. & Austin, I. (2011). *Chinese investment in Australia: Unique insights from the mining industry.* New York: Palgrave Macmillan.

Hymer, S. H. (1976). *International operation of national firms: A study of direct foreign investment.* Cambridge: MIT Press.

Johanson, J. & Vahlne, J.-E. (2009). The Uppsala internationalization process model revisited: From liability of foreignness to liability of outsidership. *Journal of International Business Studies, 40,* 1411–1431.

Klinger, P. (2013). Gas contract adds to CITIC woes. Retrieved 1 January 2014, from https://au.news.yahoo.com/thewest/a/15746940/gas-contract-adds-to-citic-woes/.

Klossek, A., Linke, B. M., & Nippa, M. (2012). Chinese enterprises in Germany: Establishment modes and strategies to mitigate the liability of foreignness. *Journal of World Business, 47*(1), 35–44.

Luo, Y., Shenkar, O., & Nyaw, M.-K. (2002). Mitigating liabilities of foreignness: Defensive versus offensive approaches. *Journal of International Management, 8*(3), 283–300.

Metallurgical Corporation of China. (2014). Measures taken by MCC to import labor to Western Australia and suggestions. In Z. Ma (Ed.), *How to invest in Australia* (pp. 123–127). Beijing: China Commerce and Trade Press.

Mezias, J. M. (2002). Identifying liabilities of foreignness and strategies to minimize their effects: The case of labor lawsuit judgments in the United States. *Strategic Management Journal, 23*(3), 229–244.

Miller, S. R., Thomas, D. E., Eden, L., & Hitt, M. (2008). Knee Deep in the Big Muddy: The Survival of Emerging Market Firms in Developed Markets. [Article]. *Management International Review (MIR), 48*(6), 645–665.

Ministry of Commerce. (2013). *2012 Statistical bulletin of China's outward foreign direct investment*. Beijing: China Statistics Press.

Ministry of Commerce. (2014). *2013 Statistical bulletin of China'' outward foreign direct investment*. Beijing: China Statistics Press.

Scott, W. R. (1995). *Institutions and organizations*. Thousand Oaks: Sage Publications.

Sofka, W., & Zimmermann, J. (2008). There's no place like home: A strategic framework to overcome liability of Foreignness in the German Car Market. *ZEW-Centre for European Economic Research Discussion Paper*. Rochester, NY.

Sun, S. L., Zhang, Y., & Chen, Z. (2013). The challenges of Chinese outward investment in developed countries: The case of CITIC Pacific's Sino Iron Project in Australia. *Thunderbird International Business Review, 55*(3), 313–321.

UNCTAD. (2014). World investment report. Retrieved 26 July 2014, from http://unctad.org/en/PublicationsLibrary/wir2014_en.pdf.

Vernon, R. (1977). *Storm over the multinational: The real issues*. Boston: Harvard University Press.

Yin, R. K. (2003). *Case study research: Design and methods*. Thousand Oaks: Sage Publications.

Zaheer, S. (1995). Overcoming the liabilities of foreignness. *Academy of Management Journal, 38*(2), 341–363.

4
Managing Post-Transaction Integration by Chinese MNCs

Introduction

Cross-border acquisition has been used by Chinese MNCs as a major approach in their OFDI in the last decade. According to the UNCTAD (2014), nearly half (49.6 per cent) of the Chinese OFDI was made through acquisitions. As discussed in chapters 2 and 3, the overall performance of Chinese OFDI in the Australian mining industries, with a few exceptions, is unsatisfactory due to their entry strategy, such as investment timing, the stage of project invested and the liability of foreignness and newness. While Chapter 3 primarily deals with the liabilities of foreignness for Chinese MNCs that adopted Greenfield as their entry strategy, this chapter focuses on the issues of post-acquisition integration for Chinese MNCs using acquisition or Brownfield strategy. Post-acquisition integration has long been regarded as one of the most important factors that influences the OFDI performance (Child et al., 2001; Hitt et al., 2001) and international acquisitions by Chinese MNCs (Williamson & Raman, 2011).

This chapter focuses on Minmetals' acquisition of a majority of assets held by an Australian listed mining company, OZ Minerals, in 2009 and its subsequent integration and development over the past five years (2009–2014). The case has been widely regarded as one of the highly successful international acquisitions by Chinese MNCs, particularly among SOEs. Since the acquisition in June 2009, Minmetals has established its Australian subsidiary Minerals and Metals Group (MMG) and has integrated and further developed it into a mid-tier global mining company over the past five years. It offers a good opportunity to examine how post-acquisition integration was implemented and what factors contributed to the acquisition performance. Before we describe and present our research, we provide a literature review in the area of international acquisition integration.

Literature review

Mergers and acquisitions (M&A) have become increasingly popular in business practice, especially international M&As that can be defined as "involving an acquirer firm and a target firm whose headquarters are located in different home countries" (Shimizu et al., 2004, p. 309). However, so far there is evidence to indicate that many M&As are unsuccessful and the success rate has shown little improvement (Marks & Mirvis, 2001, 2011; Schoenberg, 2006). M&A has been widely considered as a process that covers strategic entry thinking and target identification, transaction or acquisition, and post-acquisition integration (Hitt et al., 2001). Each stage entails a different set of factors to be considered and thus draws the attention of researchers from different disciplines. Given the complexity and importance of M&A, it is not surprising that this area has been studied by multiple disciplines from, among others, strategic, organisational and financial perspectives (Birkinshaw et al., 2000; Child et al., 2001; Haspeslagh & Jemison, 1991). Table 4.1 outlines the key issues considered and research disciplines adopted in each of the three stages of the M&A process.

In the following sections, we focus our literature review on the two stages of the M&A process: strategic thinking and target identification, and post-acquisition integration. We first briefly describe the key issues considered at the stage of strategic thinking and target identification, followed by a literature review on the post-acquisition integration.

Strategic thinking and target identification in M&A

The first stage of the M&A process is pre-acquisition preparation or "strategic thinking and target identification" (Hitt et al., 2001), which is often studied from a strategic management perspective. From the strategic management perspective, international acquisition is a crucial part of a firm's strategy either to increase its presence in the international market and acquire strategic assets or to seek resources. Therefore, how a firm's international acquisition decisions influence its acquisition performance has been a dominant theme. These decisions include the acquisition motivations, market entry modes (Greenfield, acquisition or holding) and identification of target organisations and potential strategic combination benefits between the acquired and acquiring firms.

Table 4.1 Key issues and research disciplines adopted in each of the stages of the M&A process

Stage	Strategic thinking and target identification	Transaction/ acquisition	Post-acquisition integration
Key factors considered	Motive, entry mode, target selection relatedness; planning and analysis; decision-making	Price, wealth creation, agency issues and the role of investors	Corporate control, integration approach, functional integration, cultural difference, knowledge transfer and organisational learning
Dominant discipline of research	Strategy	Finance/capital market	Organisational behaviour and culture; learning process
Performance	Effect of relatedness; achievement of corporate strategy	Stock market-based measures, financial returns	Realisation of strategic benefits of acquisition; preservation of acquired organisation's critical capability; learning and knowledge

Source: Birkinshaw et al. (2000), Child et al. (2001), Haspeslagh and Jemison (1991), Hitt et al. (2001), Javidan et al. (2004) and Larsson and Finkelstein (1999).

There are four major entry motivations: resource-seeking, strategic asset-seeking, market-seeking and efficiency-seeking. Chapter 2 has described and discussed these four motivations, thus they are not elaborated here. As discussed in Chapter 2, the driving force behind a majority of Chinese OFDI and acquisitions in Australia so far has been to seek resources, and more recently to seek markets, such as companies operating in the banking, airline and telecommunications industries. Financial gain has also frequently been observed as a motivation for Chinese acquisitions in Australia.

The manner in which firms enter international markets is another strategic decision (Cui & Jiang, 2012). This covers decisions made in favour of JVs or wholly owned subsidiaries (WOSs). Cui and Jiang (2012) found that firms are more likely to use JVs when they are under high levels of external institutional pressure to secure institutional legitimacy and alleviate institutional costs. However, a Chinese MNC opts to adopt a WOS approach when it pursues a global strategy, seeks strategic assets and encounters tough competition in the host country (Cui & Jiang, 2009). With regard to the identification and selection of target organisations, the strategic decisions are focused on the strategic benefits of the acquisitions, such as the relatedness, vertical integration or horizontal integrations, or diversification. The key question is how the international acquisition can help implement the corporate strategy of the acquired company.

Post-acquisition integration

Post-acquisition integration has long been regarded as crucial to M&A performance (Kiching, 1967) and is primarily studied from the perspectives of organisational behaviour, culture and process. The two fundamental research questions are: how acquisitions affect the organisation and employees, and how people respond to the acquisition (Javidan et al., 2004). Consequently, there are two subthemes or "groups" (Javidan et al., 2004) that focus on different issues in the post-acquisition integration. The first theme is functional integration, or "task", the "hard side" of the integration, such as marketing, production, procurement and technology, while the second theme is human or socio-cultural integration (Colman & Grøgaard, 2013), the "soft side" of the integration. Socio-cultural integration covers issues such as the differences between the acquiring and acquired organisations in culture, management style and HRM practices, or social aspects (Björkman et al., 2007). Although these two subthemes are distinct, they are intertwined and can be regarded as two sides of the same coin. The former is usually studied from the organisational perspective, while the latter from the cultural perspective.

Recognising the importance of strategic context and its impact on the degree of integration, Haspeslagh and Jemison (1991) argued that different integration approaches should be adopted. Such integration approaches can be considered as "integration strategies" (Cartwright & Schoenberg, 2006) as they provide strategic guidelines for how integration should be undertaken based on some key strategic factors (Ellis, 2004). Consequently, different types of corporate control (e.g. strategic, operational, financial) may need to be deployed based on the type of integration approach and strategic context (Child et al., 2001).

As MNCs from developing countries increasingly pursue their international acquisition, particularly in their acquisitions of organisations in developed countries, for strategic asset-seeking, over the last two decades, knowledge transfer and learning has become increasingly important for those international acquisitions. Therefore, the following subsections review and discuss four themes in the post-acquisition integration literature:

- Integration strategy (approach)
- Corporate control
- Cultural integration
- Knowledge transfer and learning

Integration strategy/approach and corporate control

The selection of the type of integration approach can be considered a strategic decision, or "integration strategy" (Birkinshaw et al., 2000) or "the choice of integration strategy" (Cartwright & Schoenberg, 2006, p. S2) in managing post-acquisition integration. Haspeslagh and Jemison (1991) have developed a typology of integration approach in their prize-winning book based on two factors: the needs for strategic interdependence and the needs for organisational autonomy for the acquired organisation. While the considerations about strategic interdependence between the acquired and acquiring firms are focused on how value is created (such as resource-sharing, functional skill transfer and general management capability), the decision on the need for organisational autonomy is often made based on how strategic capability in the acquired firms can be reserved and transferred. These four types of integration approaches are outlined in Figure 4.1.

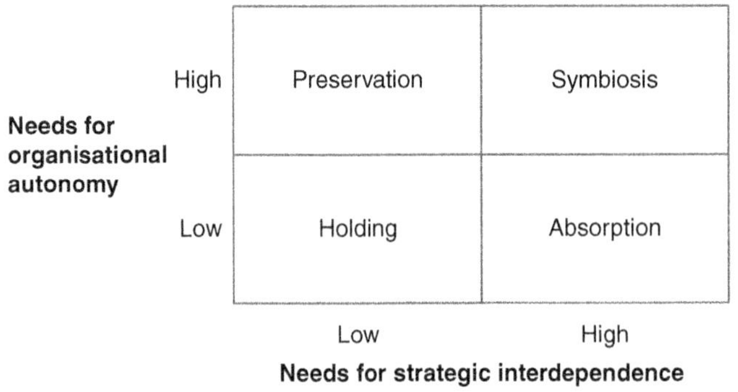

Figure 4.1 Types of acquisition integration approaches
Source: Haspeslagh and Jemison (1991).

Haspeslagh and Jemison (1991) have suggested that the preservation approach is appropriate when there is a high need for organisational autonomy and a low need for strategic interdependence. However, when strategic goals demand a high level of interdependence to generate the value expected with a low need for organisational autonomy, absorption should be considered. Symbiosis fits best with the acquisitions where both needs for strategic interdependence and organisational autonomy are high. In the case of *holding*, there is little need for strategic interdependence and organisational autonomy.

In their study of 201 acquisitions of UK companies by MNCs from the United States, United Kingdom, Germany, France and Japan, Child et al. (2001) found that any integration approach could be effective if used properly. However, the foreign acquiring company may prefer to use a particular type of integration approach. For example, American firms were more likely to use absorption, while Japanese companies often used preservation. These observations suggested that national culture and other factors could also be considered in selecting integration approach.

The research undertaken by Child et al. (2001) indicated a few exceptions in the studies of corporate control used by MNCs in their international acquisitions. Research on corporate control is limited (Shimizu et al., 2004). Effective corporate control mechanisms need to be in place to make sure that functional and cultural integrations can be implemented in order to achieve the acquisition's goals. In their study, Child et al. (2001) examined how strategic and operational controls were used by the MNCs. They found that there was a trend for MNCs to use strategic control, rather than portfolio holding, although variations existed involving strategic versus financial orientation, appointment of senior managers to the acquired firm and communications. Nevertheless, there was little correlation between integration and control. Child et al. (2001) also found that nationality was a significant factor affecting the degree of integration in addition to the quality and size of the acquired firm and the philosophy concerning subsidiary management.

Task and cultural integrations

Task and cultural integrations are critical to acquisition performance as they are mechanisms which can be used to realise strategic benefits or synergy through activity sharing and competence transfer. Task integration can be defined as "the identification and realization of operational synergies", while cultural integration as "the creation of positive attitudes towards the integration among employees on both sides" (Birkinshaw et al., 2000, p. 400). Although functional or task integration is the "value creation" (Birkinshaw, et al., 2000) or "hard" side of organisational integration, empirical studies on how it is actually implemented in organisations are surprisingly scant, compared with those on cultural integration.

Functional or task integration can take three forms: procedural, physical and managerial (Shrivastava, 1986). It may be implemented literally at every task or function between the acquired and acquiring firms, such as manufacturing, legal and environmental, technology/research and development (R&D), purchasing, finance, information system, communications, marketing and sales (Galpin & Herndon, 2014), and between the acquired firm and its suppliers or customers within its value network.

Using a survey and in-depth interviews, Birkinshaw et al. (2000) investigated the integration of R&D functions by three large Swedish MNCs. They found that the number of integration mechanisms used and problems encountered influence the level of integration achieved and the time of integration. The integration mechanisms included international staff meetings, joint R&D meetings, cultural awareness seminars, mixed project teams, R&D personnel rotation and joint R&D personnel training programmes.

To successfully implement functional integration, acquiring firms may need to develop an integration team governance model (Galpin & Herndon, 2014) that covers: (1) establishing an integration infrastructure such as the establishment of an executive team, integration project team and task forces; (2) staffing the integration project team; and (3) developing and launching the integration process. They argued that such an organisational setting can better facilitate the implementation of post-acquisition integration.

"Socio-cultural" integration (Colman & Grøgaard, 2013) becomes increasingly important in international acquisitions primarily due to the rapid development of globalisation over the last two decades, especially the rapid growth of MNCs from developing countries and their international expansions (UNCTAD, 2014).

The challenges for cultural integration in international acquisitions come from the cultural differences at both national and organisational levels. At the national level, differences in institutions, philosophies and cultural values (Barkema et al., 2015) influence the way of conducting business operations (Scott, 2014). Moreover, the differences in institutional logics (Thornton et al., 2012) often produce international complexity and thus differences in managerial practices, thereby making the cultural integration more challengeable.

At the organisational level, corporate culture can also be very different, particularly between companies from different countries. Such cultural differences at the corporate level can adversely affect cooperation of managers between the acquired and acquiring countries in post-acquisition integration (Weber & Shenkar, 1996), such as linguistic and cultural problems in communications (Child et al., 2001).

The cultural differences at the national and corporate levels create many challenges in international acquisition integration as a "double-layered acculturation" (Barkema et al., 1996) and this might create problems for the full realisation of integration potential.

Learning and knowledge

Organisational learning can be defined as "the transfer of an organisation's experience from one event to a subsequent one" (Barkema & Schijven, 2008, p. 596). Acquisition learning can be undertaken in three ways: (1) from one's own experience; (2) from others' experience; and (3) from the knowledge embedded in the acquired firm.

Although learning has long been studied in M&A literature (Pennings et al., 1994), the importance of learning and capability development in international acquisitions has been increasingly recognised by MNCs and thus has been the subject of greater research recently. Early research (1970s–mid 1990s) on acquisition learning was mainly theoretical and primarily based on the observation of the experience curve. A key focus was on the relationship between an organisation's acquisitions experience, often measured by the number of acquisitions prior to the focal acquisition studied, and the focal acquisition performance. As explicitly pointed out by Barkema and Schijven (2008), there are three assumptions in the use of the experience curve: (1) experience is always positive to learning; (2) experience automatically leads to learning; and (3) firms only learn from their own activities. Consequently, early studies on the relationship between acquisition experience and performance, coupled with problems in their experience measurements, were often inconclusive or even controversial.

In recognising these three assumptions based on the experience curve, recent studies on acquisition learning have relaxed these suppositions and have concentrated on the diversity (homogeneity vs. heterogeneity) and relevance of experience. Attention has also been given to organisational conduits and mechanisms that can be used more effectively and efficiently to convert experience to learning and to learn from others.

With regard to the nature of experience, the relevance of prior acquisition experience has been a focus of the literature in terms of region, country and industry. In general, the higher the relevance of prior acquisition experience, the higher the success in the subsequent acquisitions. However, the impact of the degree of homogeneity of experience on acquisition performance is still inconclusive. Homogeneous experience can be more instrumental in exploiting subsequent similar acquisitions, while heterogeneous experience can be more useful for exploring acquisitions in other different acquisition contexts, such as different industry or host country.

To convert an organisation's own experience and observation into learning and capability, effective mechanisms should be in place. These mechanisms, or "deliberate learning mechanisms" (Zollo & Winter, 2002), include active collection and evaluation of performance feedback on acquisitions (Haleblian et al., 2006), transfer of expatriates from the acquiring firm to the acquired one (Hébert et al., 2005), knowledge management (experience articulation and codification) (Zollo & Singh, 2004), establishment of an organisational function responsible for acquiring and accumulating experience (Kale & Singh, 2007), or even learning from advisers (Kim et al., 2011). Regarding the conduits of learning, deploying expatriates to the acquired firm is very effective in enhancing organisational learning.

Based on the discussion above, the relationship between learning sources, mechanisms, learning and acquisition performance can be depicted in Figure 4.2.

Failure to implement post-acquisition integration by Chinese MNCs has been regarded as a key factor influencing their international acquisition performance (Williamson & Raman, 2011). However, there is little research on the post-acquisition integration by MNCs (Deng, 2012, 2013), with a few exceptions; for example, Liu and Woywode (2013) used case studies of acquisitions by Chinese MNCs in Germany and found that Chinese investors adopted a new approach called "light-touch". This method combines "preservation" and "symbiosis" (Haspeslagh & Jemison, 1991) based on the contextual characteristics and need for learning from the acquired German firm. The researchers also reported that such a light touch approach can be very beneficial to both Chinese acquiring firms and German acquired companies. Deng (2010) argued that the absorptive capacity of Chinese acquiring firms is critical to the success of their international acquisitions. Such absorptive capacity has three dimensions: ability to understand, integrate and apply strategic assets and their key determinants. Nevertheless, Deng

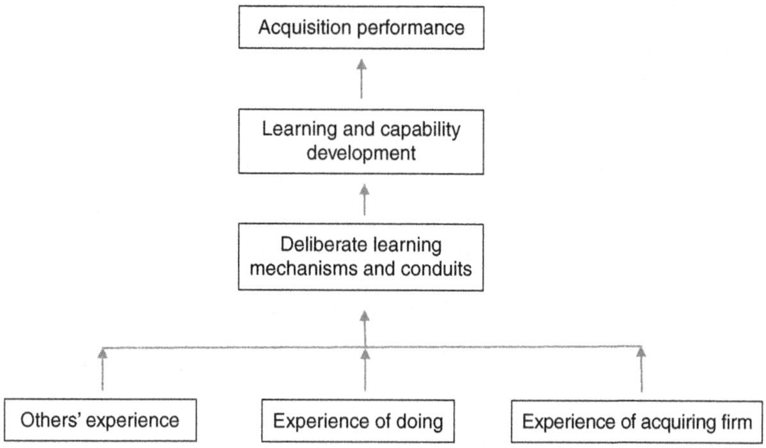

Figure 4.2 Organisation learning in acquisition: Sources, mechanisms and performance

only used secondary data regarding a failed acquisition, namely TCL's acquisition of a French firm, Thomson, to illustrate how the lack of absorptive capacity in TCL was a main cause for acquisition failure.

In summary, given that international acquisition is "a complex economic, social and political activity" (Bower, 2004), a variety of factors have been identified in the existing literature on post-acquisition integrations. Adopting the antecedents-strategy-performance framework proposed by Peng (2006), we have developed a theoretical framework in Figure 4.3 and will use this in our study of post-acquisition integration by Chinese MNCs.

Chinese MNCs are mainly state-owned enterprises (SOEs) and are latecomers in international acquisitions (Deng, 2013). These SOEs have played a dominant role in China's economy (Ralston et al., 2006) and have undertaken the majority of international acquisitions (Ministry of Commerce, 2014). However, Chinese SOEs lack firm-specific advantages (FSAs) and international experience and are often subject to China's institutional and government intervention (Luo et al., 2010), which can be a double-edged sword for these Chinese SOEs (Gao et al., 2015). Thus, the post-acquisition integration approach taken by Chinese SOEs may be different from MNCs in developed countries. The following case study focuses on the key question of this chapter, namely, how Chinese SOEs undertake integration in their international acquisitions.

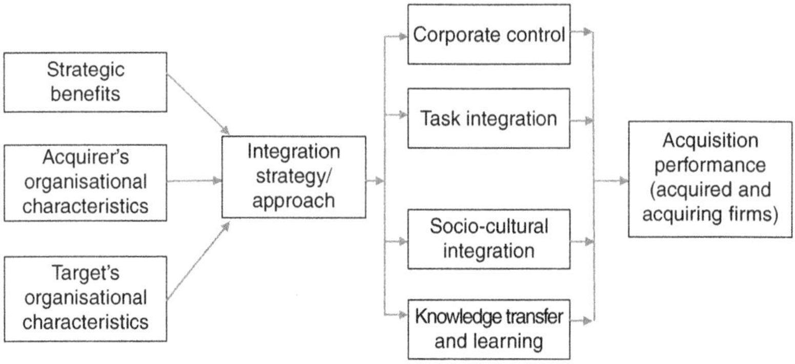

Figure 4.3 The effect of after-acquisition integration on acquisition performance
Source: Adapted from Peng (2006) and Birkinshaw et al. (2000).

Research methods

Research design

This chapter focuses on the post-acquisition integration by Chinese MNCs (who are mainly SOEs) and is exploratory in nature, given that post-acquisition integration is a relatively new phenomenon in these Chinese MNCs. Case study methodology is very appropriate for such a research purpose (Yin, 2003) and was employed in the present study. By the end of 2013, Australia became the fifth largest destination for Chinese OFDI, with 60.6 per cent of this investment in the mining industry (Ministry of Commerce, 2014). After-acquisition integration is widely regarded as a complex and key managerial task (Child et al., 2001; Galpin & Herndon, 2014; Hitt et al., 2001). Therefore, it is appropriate to use case studies to explore key issues and their effect on post-acquisition performance.

Sampling consideration and data collection

Minmetals is one of the largest Chinese MNCs pursuing its international expansion strategy. It acquired a majority of the assets in OZ Minerals in 2009 and subsequently established its subsidiary (MMG) in Australia after completing the acquisition. Over the past five years, Minmetals has been very active in integrating the company it acquired (MMG). Minmetals has a relatively long history of implementing post-acquisition integration and thus offers much information on its integration process and the key issues in conducting this integration.

A combination of data collection was conducted for this study: secondary data collection and interviews with senior managers at both Minmetals and MMG. An extensive secondary data collection was first conducted from multiple sources: media, company websites and the Internet. While each source had its own focus, the materials collected from each source were also used for cross-validating or supplementing information from the other two sources.

Factiva is an e-database offering the most comprehensive coverage of newspaper publications. M&As, particularly large international ones, have been widely regarded as major corporate activities, thereby attracting much attention from media, regulators and the public in both host and home countries. The media we used provided extensive coverage of this particular acquisition and its after-acquisition events due to the acquisition's size and ownership by a major Chinese SOE. We used Minmetals and MMG as key words in searching two leading newspapers included in Factiva: *The Australian*, which is published by News Corporation, and *Australian Financial Review*, which is the leading Australian business newspaper and is published by the Fairfax Media Group. We studied these two newspapers from January 2008 to the end of December 2014.

The Minmetals and MMG websites were also searched to access their sustainability reports, annual reports, announcements and investor presentations.

Internet searches focused on reports by other media on the after-acquisition integration and further development of MMG after its establishment. The Internet supplemented information identified by the other two sources on Minmetals' after-acquisition integration. For example, some major historical events relating to Minmetals, particularly those occurring before 2008 and taking place outside Australia, were not well covered by Australian newspapers, although they were briefly mentioned on the Minmetals website. In addition, a leading Chinese search engine, *Baidu*, was used for collecting Chinese media reports on relevant information about Minmetals' acquisition and its after-acquisition integration process.

Interviews were conducted at both Minmetals and MMG with four senior executives (three at Minmetals' headquarters [EA, EB and EC] and one at MMG [ED]) who were heavily involved in the acquisition preparation, after-acquisition integration and further development. These interviews were conducted from late 2012 to early 2015 as part of a large and ongoing research project. Where permitted, the interviews were digitally recorded; in other cases, notes were taken during the interviews.

The interviews lasted for 60 to 120 minutes. All the recorded interviews were transcribed.

Data analysis

The titles of newspaper articles found on Factiva were first read independently by two researchers for relevance to this study. If relevant, the full article was downloaded from Factiva for further analysis. Factiva contains the full texts of *the Australian* and *Australian Financial Review*. The articles were analysed for: (1) compiling a chronological events of Minmetals from 2008 to 2014; and (2) supplementing the after-acquisition practices reported by Minmetals in its annual reports and other relevant materials downloaded from its website and MMG's website.

The materials downloaded from the websites of Minmetals and MMG, particularly their annual reports, were analysed to identify the company's major issues and approach in conducting its post-acquisition integration.

The data collected from interviews were transcribed and analysed in chronological order first and then by theme, and these were displayed with a "time-ordered matrix" (Miles & Huberman, 1994). We also used a "role-ordered matrix" (Miles & Huberman, 1994) to classify our findings based on the role of interviewees in order to generate a more comprehensive picture about the after-acquisition integration process.

Background information about Minmetals and MMG

Minmetals is one of 112 SOEs under the direct control of the Chinese central government (State-owned Assets Supervision and Administration Commission [SASAC], 2015). The company was established in 1950 as a major import and export company for minerals and metals, particularly non-ferrous metals, and was originally named China National Metals & Minerals Import & Export Corporation. It was renamed China Minmetals Corporation in January 2004.

Since the implementation of China's economic reform in 1979, particularly over the past two decades, Minmetals has been expanding through M&A. Currently, Minmetals primarily engages in exploration, mining, smelting, processing and trading for metals and minerals, and it has six business centres or divisions: non-ferrous metals centre, iron ore and steel business division, ferrous mining business division, finance services division, real estate business division and science and technology division. In 2013, Minmetals' revenue and net profit were US$68.5 billion and US$1.07 billion respectively. It was ranked 192nd among the Fortune Global 500 (China Minmetals Corporation, 2014).

MMG was established on 17 June 2009 after Minmetals completed its acquisition of the majority of assets of OZ Minerals for US$1.386 billion. MMG had four mines in operation at the time of completion of the acquisition; three in Australia (Century in Queensland, producing about 500,000 tons of zinc concentrate annually; Golden Grove in Western Australia, mining zinc, copper and lead; Roseberry in Tasmania producing zinc, lead and gold); and one operation in Laos (Sepon) producing gold and copper.

Prior to being acquired by Minmetals, the former OZ Minerals was the merged entity of Oxiana and Zenifex established on 1 July 2008, just before the GFC, with a market capitalisation of approximately A$12 billion at its peak, making it the third largest diversified mining company listed on the ASX (OZ Minerals, 2008b). However, by the end of November 2008, it had a total and net debt of A$693 million and A$434 million respectively. Worse still was the fact that A$560 million of its debt matured at the end of January 2009, even after an extension was granted (OZ Minerals, 2008a). Apart from financing, the former OZ Minerals was robust in every other aspect, such as management, mining operations, exploration and marketing. The strong management at OZ Mineral can be illustrated by the fact that Oxiana was one of the listed companies with the best total shareholder return between 2001 and 2006. The banks' refusal to refinance the former OZ Minerals caused the company's forced sale.

Finding and discussion

Integration strategy and corporate control

Two major decisions faced Minmetals during the acquisition stage: how to control MMG and to what degree to integrate it into Minmetals' business. Corporate control is very important as it can have widespread and substantial implications for post-acquisition integration. These two decisions were intertwined and influenced by China's institutions and Minmetals' human resources and corporate strategy.

Institution considerations

The most important factor considered in relation to corporate control was the impact of China's institutions and corporate identity. As a large

Chinese SOE, having a majority control in its international acquisition was consistent with its organisational identity. Although having a majority control was deemed desirable and necessary, the challenge for Minmetals was how to control this ownership, as was expressed by a senior executive (EA) with regard to Minmetals' acquisition strategy and control as a large Chinese SOE:

> [We are] "looking for middle and high quality of mineral assets and mining companies. We would like to acquire some of them... and have a majority ownership to control them".

He then continued elaborating on why a majority ownership control is necessary: "We (Minmetals) are a centrally-controlled SOE. If you (we) do not have a majority control, why you go there."

The decision on the corporate control model for MMG was taken with due consideration of Minmetals as a SOE and of its corporate strategy. As EA described:

> This [MMG] should be Chinese controlled and Westerners managed, a combination of West and East. We [Minmetals] cannot manage it. We acquired this mining company [MMG]. What I would like to control is only strategy.

As one of the Chinese centrally controlled SOEs, Minmetals is supervised by the SASAC. Direct government intervention in Chinese SOEs has been widely reported (Gao et al., 2015; Luo et al., 2010). Moreover, MMG was considered as a Tier-2 subsidiary within Minmetals, as it was part of the Non-ferrous Metals Division within Minmetals. Thus it could be subject to potential formal or informal interferences from within Minmetals. This could adversely affect MMG's management and operations and be detrimental to the managerial relationship between Minmetals and MMG.

In order to exercise its strategic control, a new MMG board was established shortly after the acquisition was completed, with four inside directors, two Australian executive members and three independent members. The preservation approach was adopted by Minmetals in its post-acquisition integration. Therefore, MMG was managed as an independent subsidiary.

The MMG's board has played two important roles: exercising strategic control over MMG and buffering the potential intervention from the Chinese government and within other parts of Minmetals. This model can be summarised as "Chinese (company) holds a majority of

MMG's shares, Westerners (managers) manages its operations, and the board controls it" (EC). Such a control mechanism can help Australian management team run business operations with their familiar Western management style and cultural climate on the one hand, and on the other hand, the board is able to concentrate on the key issues that have major impact on the corporate development. This is well elaborated by one interviewee (EB):

> The board is responsible for making major MMG's strategic decisions, such as strategy development, major investment, budget approval, major external financing, and appointment of executives to MMG, while the management team at MMG is responsible for its operations.

To facilitate the organisational integration and alleviate the potential intervention from within Minmetals in MMG, the President of Minmetals took the position of Chairman of the board. The board met six times a year to make strategic decisions regarding MMG. One Chinese expatriate was appointed in early 2010 as Board Support Officer to liaise between the inside board directors and the management at MMG. He was the only expatriate sent by Minmetals to MMG until July 2013 when another expatriate was sent to MMG to be responsible for liaising with Chinese banks for MMG's further development. Table 4.2 provides a chronological list of major events in the acquisition and post-acquisition development.

Human resource considerations

How to control MMG was a major challenge for Minmetals. This issue was questioned by Chinese senior government officials who wanted to know whether Minmetals could successfully manage MMG after its acquisition. The issue of corporate control was hotly debated as one executive (EC) described it: "There were two different views at Minmetals at the time of acquisition: deploying expatriates to control it or keeping the existing management team to run the business." Nevertheless, lack of internal competent executives was another key reason for adopting the latter as one executive (EA) commented:

> Two things are important in cross-board acquisition: experience and management team. If you don't have a management team, how you are going to manage the assets acquired overseas unless you would use the existing management team. In the case of MMG, I would like to use Western management expertise because we haven't a good management team (to manage it).

Table 4.2 A chronological list of major events in Minmetals' acquisition and post-acquisition development

Date	Event
	Negotiation period from 29 November 2008 to 15 February 2009
28 November 2008	OZ Minerals requested to halt trading.
	Transaction period from 16 February to 11 June 2009
16 February 2009	Minmetals proposed 100% acquisition of OZ Minerals for US$1.84 billion (A$2.6 billion).
27 March 2009	Federal Treasurer Wayne Swan rejected the Minmetals' original takeover bid and requested Minmetals to revise its bid application.
30 March 2009	Minmetals submitted a revised offer of US$1.2 billion (A$1.7 billion) for OZ Minerals.
23 April 2009	The Australian government approved the proposed majority acquisition of OZ Minerals by China Minmetals but imposed a number of strict conditions on the sale.
6 May 2009	Former Managing Director of OZ Minerals agreed to stay after its acquisition by Minmetals.
10 June 2009	Minmetals increased its offer by 15% to US$1.386 billion (A$1.96 billion).
11 June 2009	OZ Minerals' annual general meeting approved the deal.
16 June 2009	Minmetals completed its acquisition.
	Post-transaction integration and management (from 18 June 2009 to 31 December 2010)
18 June 2009	Minerals and Metals Group (MMG) was established.
22 March 2010	MMG reported a net profit of US$180.7 million for 2009, covering the first seven months since it was established.
31 December 2010	Hong Kong listed Minmetals Resources (MMR) bought Minerals and Metals Group (MMG) from its parent company for US$1.85 billion (A$1.87 billion).
	Expansion since 1 January 2011
26 January 2011	Plan to invest US$700 million in its zinc mine in Queensland.
4 April 2011	MMR bid and for listed copper miner Equinox Minerals, which was mining in Zambia and Saudi Arabia, for C$6.3 billion (C$7.00/share).
26 April 2011	Barrick Gold Corporation trumped MMR for US$7.15 billion (C$7.3 billion C$8.15/share) for Equinox Minerals.

24 June 2011	MMG put the Tasmanian nickel mine, Avebury, up for sale due to its small scale. Avebury was not in operation even before it was acquired by Minmetals as part of the former OZ Minerals.
7 September 2011	MMR took a placement of 4,000,000 shares at an issue price of A$1.25 in Havilah Resources to partner on exploration in South Australia and agreed to spend A$12 million in the next five years.
16 September 2011	MMR sold its trading and fabrication assets to parent company China Minmetals Nonferrous for US$726.8 million (A$713 million); was left with only the assets it bought from OZ Minerals in 2009 and a small bauxite interest.
1 October 2011	MMR bid for Anvil Mining with its interests in the copper area of Katanga in the Democratic Republic of Congo for C$1.33 billion (C$8.00/share), closed in February 2012 with over 90 per cent shared secured.
28 October 2011	Minmetals launched its Australian sustainability report.
20 March 2012	MMG acquired Avil Mining.
27 August 2012	Minmetals Resources changed its registered English company name to MMG Limited.
17 December 2012	Dugald River Project approved by the Board, budgeting for A$1,488 million.
1 December 2013	MMG ceased gold operations at its Sepon mine in Laos.
31 July 2014	MMG completed the acquisition of Las Bambas as part of a joint venture (62.5 per cent) for US$7.005 billion.

Note: A$ – Australian dollar (A$1 = US$0.705 in early 2009); C$ – Canadian dollar.

Another interviewee (EC) commented on this issue further:

> We were considering whom we can possibly send to Australia for managing MMG. At that time, we could have at most 2 potential managers to be sent based on their managerial experience compared with that of existing management team, not to mention about social skills.

Minmetals' corporate strategy and the role of MMG

Developing MMG as a platform for international expansion was part of Minmetals' growth strategy. As discussed above, Minmetals was established as a trading company. It has been transformed from a trading company to a conglomerate using backward vertical integration since

2004. Consequently, Minmetals' total annual revenue increased from RMB156 billion in 2007 to RMB414.6 billion in 2013, an annual compound growth rate of 17.7 per cent. As claimed by EC, the acquisition of a majority of the assets held by OZ Minerals offered an excellent opportunity for Minmetals to implement a growth strategy, as it had a very competent management team with a complete "organising system".

The adoption of a "preservation" approach and a strategic control board appears to be very appropriate for Minmetals' post-acquisition integration. The performance of such a corporate control structure and integration strategy so far shows that they are very effective in achieving Minmetals' control intention and preservation of MMG's capability for its international expansion. At the same time, it also motivates the management team to actively develop its initiatives, thereby meeting both requirements of strategic control and operations. The quality of the management team in the target company is a necessary condition for adopting such an integration strategy and corporate control approach. As discussed in the section above, the acquired firm was very competent in management and mining operations before the acquisition.

Organisational integration

Functional integration is very important to the realisation of potential acquisition benefits (Haspeslagh & Jemison, 1991). However, the institutional setting of Minmetals as a China's centrally controlled SOE under the administration and supervision of the SASAC requires some unique organisational integration, particularly in the area of procedural integration.

Compliance with the SASAC regulations and requirements was the first priority for organisational integration at MMG. SASAC has many specific administrative regulations and reporting requirements. For example, major investments, number of employees and amount of salary have to be reported regularly to SASAC. Major safety accidents have to be reported to Beijing within two hours of their occurrence.

Realising the importance of these procedural and functional integrations, Minmetals set up a MMG Management Integration Team after the completion of the transaction. This team included managers and employees from departments such as finance, planning, investment, legal, HR, safety and environmental protection and risk management, as well as several external consulting firms. A team consisting of ten functional managers from Minmetals' Non-ferrous Division was sent to MMG in late 2009 for two weeks to communicate in detail the requirements needed to integrate MMG's relevant functions into the Minmetals' system. Consequently, a number of joint systems were established, such as a budget management system, an accounting and statement system and an investment management system (China Minmetals Corporation, 2011, p. 37).

From MMG's perspective, probably the most value-added activities are support and assistance from Minmetals for MMG's procurement, sales and finance based on Minmetals' strengths and capability. As a large Chinese SOE, Minmetals has developed its solid relationship with Chinese banks. China has accumulated a high level of foreign currency reserves – nearly US$4 trillion in 2014. Thus, Minmetals is able to secure loans from Chinese banks under favourable conditions, often in terms of lower interest rate and long-term loans. This was clearly expressed by a senior manager (EA) during the interview. These benefits ensure MMG has a good cash liquidity and low financing costs, thereby helping improve MMG's financial performance. Moreover, Minmetals has assisted MMG in procuring raw and auxiliary materials, using its expertise and network in China in identifying potential suppliers, helping to negotiate prices, dealing with Chinese customs and tracking the progress of orders, such as mining equipment and machinery. This has substantially helped MMG reduce its procurement costs. Similarly, Minmetals can also help MMG sell its minerals in China if needed due to excess production, although MMG has signed off-take agreements with several international buyers for its mineral products. Figure 4.4 shows the initiatives taken by Minmetals in its organisational integration with MMG after the acquisition.

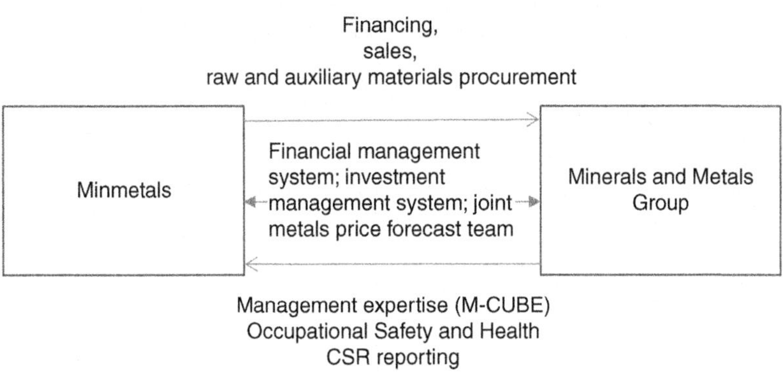

Figure 4.4 Functional integration between Minmetals and MMG

Cultural integration

Although "the preservation" approach (Haspeslagh & Jemison, 1991) was used by Minmetals for the post-acquisition of MMG, which is geographically separated from Minmetals, there was still a strong need for cultural integration. Better coordination and cooperation was needed in MMG's management and operations. A number of initiatives have been taken even before the transaction was completed. For example, three senior managers from the Non-ferrous Division were sent to Melbourne to talk to every member in the executive team of OZ Minerals before the transaction was completed. This one-on-one conversation aimed to communicate to Australian managers Minmetals' vision and strategy and to establish how it would develop the target company after the acquisition. This helped them understand Minmetals' vision and culture. As a result, the entire executive team was retained after the acquisition.

Shortly after the acquisition, a Chinese cultural training programme was organised for all MMG's frontline and middle managers. It was conducted by The Confucius Institute in Melbourne. This half-day programme aimed to improve their understanding of Chinese values, business behaviour and etiquette.

Another important method of cultural integration has been mutual participation in events at both Minmetals and MMG. For example, MMG's executives are invited to attend Minmetals' annual conference in Beijing, while Minmetals' senior managers are invited by MMG to attend its annual executive team CEO Conference. Moreover, several Minmetals' managers have been invited to participate in MMG's management training programme. These initiatives have helped mutual understanding of Chinese and Australian cultures.

Finally, an internship programme was initiated in 2012 in which two or three young employees from Minmetals are sent to MMG for three to six months every year. Under this internship programme, the interns learn not only business skills but also the Australian culture. Since 2014, MMG has also sent its employees (usually two or three staff annually) to Minmetals for one year to understand how business is conducted at Minmetals and to learn the Chinese culture.

Knowledge transfer and learning

The importance of knowledge transfer from MMG to Minmetals and learning was fully recognised by Minmetals even before the transaction was completed, as described by one executive (EA): "I would like to have two resources from this acquisition: mineral resource and human resource."

The acquisition and retaining of all executives of OZ Minerals have provided many opportunities for Minmetals to learn from MMG and transfer the knowledge from MMG to Minmetals. One interviewee (EC) described such opportunities as follows:

> MMG was an international mining company with a full comprehensive and integrative management and organisational system. It is an organising system. This is very difficult to develop and extremely valuable to us [Minmetals]. It is not simply a mode or a function.

Functionally, there are many examples of such knowledge and learning, he continued:

> Minmetals has learned a lot from MMG. One example is the decision-making process. You know, decision-making in Chinese firms is usually made by the top manager and often based on the government authority. However, decisions in MMG are made based on the facts and data. Another example is the implementation of occupational safety and environmental protection. In China, implementation is always problematic. MMG has a comprehensive set of policies and operating system for these. Moreover, in mineral prospecting, Minmetals has learned how to use faming, rather than JV in dealing with partners for risk management.

Recognising the huge potential for knowledge transfer and learning, Minmetals launched its M-CUBE (or M^3, Minmetals, MMG and Management) project in early 2010 (China Minmetals Corporation, 2011) shortly after acquisition. The project aimed to learn from MMG for the further enhancement of management at Minmetals. The team consisted of managers from six functional areas: employee training, strategic planning, financial management, risk management, work safety management and supervision of new engineering projects. A key purpose was to exchange ideas with the relevant functional managers at MMG. The whole team was sent to MMG's Headquarters in Melbourne in mid-July 2010 in order to learn MMG's management techniques. MMG provided strong support and cooperation for this project with participating senior managers from its Business Support Unit throughout the project. As a result, more than 50 managerial tasks in 13 functional areas were identified for improvement at Minmetals, including risk assessment, project audit, reporting system, cost management, security system and training (China Minmetals Corporation, 2011).

Even over five years after the acquisition of MMG in 2009, Minmetals is still learning. In recent years, several senior staff members from MMG's Communications Office were sent to Minmetals to assist its development of annual sustainability reporting, as discussed by a Senior Manager (ED) at MMG. This is also confirmed by one interviewee (EC) who stated: "We (Minmetals) are still learning from MMG even now five years after the acquisition."

MMG's performance

Since the acquisition in 2009, MMG's financial performance has been reasonable compared with other international competitors. As a mining company, MMG's performance is heavily impacted by the changes in the global minerals prices. As shown in Table 4.3, earnings per share varied, peaking in 2012, and dramatically declining since 2013 due to the drop in the global metals price. For example, the copper price dropped from about US$3.75 per ounce in early 2013 to US$2.5 in late 2014 (London Metal Exchange, 2015).

From a growth perspective, MMG has been growing since 2009. Over the past five years, its revenue increased from US$1.6 billion in 2009 to US$2.48 billion in 2014, an increase of over 50 per cent in revenue over the five years within the context of a substantial drop in global metal prices. After its acquisitions in Congo and Peru, MMG's total assets climbed to US$13.49 billion, more than 300 per cent increase in five years. Its rapid growth was driven by Minmetals' growth strategy as both acquisitions were funded by Chinese banks with a strong support

Table 4.3 The financial performance of MMG between 2009 and 2014

(Million US dollar)	2009	2010	2011	2012	2013	2014
Revenue	1,649.7	3,582.1	2,228.30	2,499.4	2,469.8	2,479.8
EBITA	242.4	560	1,063.80	853.2	750.9	780.8
EBIT	220.3	560.1	755.3	405.6	278.3	243.7
Underlining profit	224.8	430.4	483.6	217.5	122.5	99.2
Earnings per share (%)	6.18	9.04	3.64	11.90	1.95	1.96
Total assets	3,098.2	3,466.9	3,453.5	4,659	4,683.5	13,490
Total liabilities	1,886.1	2,933.5	1,959.1	2,973	2,866.7	10,515.4
Total equity	1,212.1	533.4	1,494.4	1,686	1,816.8	2,974.6

Sources: MMG's annual report (2010, 2011, 2012, 2013, 2014).

and the direct involvement of Minmetals in securing their financing. These acquisitions were strongly supported by China's central government because of MMG's acquisition competence and its satisfactory performance record after the purchase.

Conclusion and implications

This chapter aims to explore how Chinese MNCs implement their post-acquisition integration. Using a case study method, we investigated how Minmetals conducted its integration after it acquired a majority of the assets held by OZ Minerals in 2009. The case provides an excellent research opportunity five years since the acquisition. Several lessons can be drawn from Minmetals' after-acquisition integration of MMG based on the findings presented and discussed.

Acquiring a local, well-managed company in a developed country is a quick way for Chinese MNCs to catch up with their international competitors and further international expansion. Chinese MNCs not only lack international business experience but also suffer liabilities of foreignness (LOF), as described and analysed in Chapter 3. Thus, such an acquisition is a very effective way for Chinese MNCs to overcome their LOF and their comparative disadvantage to operating in developed countries.

Chinese institutions affected how the post-acquisition was conducted by Minmetals, particularly in terms of the corporate control adopted. As a large Chinese MNC, having a majority control over its acquired firm appears to have been a dominant logic for Minmetals. Corporate control is probably one of the most important issues for after-acquisition integration. It is also the most challenging issue for Chinese MNCs. Both control structure and approach are important. Structurally, Minmetals established a board and adopted a strategic control approach to exercise its control over MMG. This approach is very appropriate within this acquisition context as MMG was expected to be a platform for Minmetals' further international expansion and was managed by an Australian team. Besides strategic control, another key role played by the board is to buffer the potential formal or informal interferences from the Chinese government and other leaders within Minmetals.

The integration approach adopted, combined with from the strategy of not deploying expatriates to executive positions, was adopted by Minmetals for two main reasons: preservation of MMG's capability for learning and further international expansion, and lack of competent executives from Minmetals to manage MMG. This approach is also consistent with Minmetals' corporate growth strategy and proved to be successful.

Organisational integration was still needed even if the "preservation" approach was adopted. One important lesson from this case study is that Chinese MNCs are active in exploring the cooperation and coordination of value chain activities to reduce procurement costs or increase sales in China through leveraging the acquiring company's expertise and capability in the Chinese market. They are also applying conventional ways of sharing activities and transferring competences between the acquired and acquiring companies for synergy.

A further lesson drawn from this case study is that learning was also important to Minmetals' acquisition. A long-term approach to knowledge transfer from MMG was used by Minmetals. Acquiring a relatively large Australian company and retaining all senior managers offered considerable potential for reverse knowledge transfers, both technically and managerially. Knowledge and skills at organisational, functional and task levels can be potential targets for transferring between subsidiaries and headquarters.

Finally, leadership is critical to the after-acquisition integration implementation in this case. The adoption of a "preservation" approach with no expatriates appointed by Minmetals to senior managerial positions at the acquired firm was innovative and challenging from a large SOE's perspective. The vision and support from top leaders in the acquired firm is extremely important to the success of post-acquisition integration.

Several implications can be proposed based on these lessons. First, a systematic integration plan is important to the successful implementation of post-acquisition integration. Such a plan needs to be guided by the acquirer's strategy, its internal resources and competence, and the role of the acquiring company in the future. Issues which need to be planned include corporate control structure and mechanisms (e.g. appointment of senior management and board members), organisational integration, human resources and cultural integration, and knowledge transfer between the acquiring and acquired companies.

Second, trust is a key issue in corporate control and knowledge transfer. Although effective corporate control mechanisms should be in place, trust at the individual level is also critical for effective management. A high level of trust between the senior managers of the acquiring firm and executives at the acquired company should be developed to ensure that the acquisition objectives can be achieved and business at the acquired firm can be operated effectively and efficiently. Similarly, trust is also needed in knowledge transfer. Building trust at both the individual and organisational levels is time-consuming.

Third, intangible assets can be as valuable and important as tangible assets in international acquisitions. Similarly, managerial competencies can be as valuable as technical ones. Often the target firm is evaluated by its price-earnings ratio (P/E). Intangible assets, such as the organisation's systems, policies, culture and standards are not a crucial part of the due diligence in acquisitions, and if considered, are difficult to assess. As discussed in this chapter, the retained Australian management team was very competent and has played an indispensable role for Minmetals' further international expansion. Therefore, intangible assets should be considered and evaluated when conducting the due diligence for an acquisition.

Finally, human resources are a critical factor that influences not only every stage of the acquisition process, but also the further development and performance of the acquired firm. Chinese MNCs need to develop their human resources to exert effective and efficient control over the acquired firms and at the same time to learn to recruit talents from international markets.

References

Barkema, H. G., Bell, J. H. J., & Pennings, J. M. (1996). Foreign entry, cultural barriers, and learning. *Strategic Management Journal, 17*(2), 151–166.

Barkema, H. G. & Schijven, M. (2008). How do firms learn to make acquisitions? A review of past research and an agenda for the future. *Journal of Management, 34*(3), 594–634.

Barkema, H. G., Xiao-Ping, C., George, G., Yadong, L. U. O., & Tsui, A. S. (2015). West meets east: New concepts and theories, *Academy of Management Journal, 58*(2), 460–479. doi:10.5465/amj.2015.4021.

Birkinshaw, J., Bresman, H., & Håkanson, L. (2000). Managing the post-acquisition integration process: How the human integration and task integration processes interact to foster value creation. *Journal of Management Studies, 37*(3), 395–425.

Björkman, I., Stahl, G. K., & Vaara, E. (2007). Cultural differences and capability transfer in cross-border acquisitions: the mediating roles of capability complementarity, absorptive capacity, and social integration. [Article]. *Journal of International Business Studies, 38*(4), 658–672.

Bower, J. (2004). When we study M&A, what are we learning? In A. L. Pablo & M. Javidan (Eds.), *Mergers and acquisitions* (pp. 235–244). Oxford: Blackwell Publishing.

Cartwright, S. & Schoenberg, R. (2006). Thirty years of mergers and acquisitions research: Recent advances and future opportunities. *British Journal of Management, 17*, S1–S5. doi:10.1111/j.1467-8551.2006.00475.x

Child, J., Falkner, D., & Pitkehly, R. (2001). *The management of international acquisitions*. Oxford: Oxford University Press.

China Minmetals Corporation. (2011). *2010 China Minmetals Corporation sustainability Australian report*. Beijing: China Minmetals Corporation.

China Minmetals Corporation. (2014). Sustainability Report of Minmetals 2013. Retrieved 1 January 2015, from http://www.minmetals.com/english/SocialContribution/

Colman, H. L. & Grøgaard, B. (2013). Integration vacuum: Creating action space for global strategy implementation in international acquisitions. *Thunderbird International Business Review, 55*(4), 405–418. doi:10.1002/tie.21553

Cui, L. & Jiang, F. (2009). FDI entry mode choice of Chinese firms: A strategic behavior perspective. *Journal of World Business, 44*(4), 434–444.

Cui, L. & Jiang, F. (2012). State ownership effect on firms' FDI ownership decisions under institutional pressure: A study of Chinese outward-investing firms. *Journal of International Business Studies, 43*(3), 264–284. doi:10.1057/jibs.2012.1

Deng, P. (2010). What determines performance of cross-border M&As by Chinese companies? An absorptive capacity perspective. *Thunderbird International Business Review, 52*(6), 509.

Deng, P. (2012). The internationalization of Chinese firms: A critical review and future research. *International Journal of Management Reviews, 14*, 408–427.

Deng, P. (2013). Chinese outward direct investment research: Theoretical integration and recommendations. *Management & Organization Review, 9*(3), 513–539. doi:10.1111/more.12030

Ellis, K. M. (2004). Managing the acquisition processes: Do differences actually exist across integration approaches? In A. L. Pablo & M. Javidan (Eds.), *Mergers and acquisitions: Creating integrative knowledge* (pp. 113–132). Oxford: Blackwell Publishing.

Galpin, T. J. & Herndon, M. (2014). *The complete guide to mergers and acquisitions: Process tools to support M&A integration at every level* (3rd ed.). San Francisco: Jossey-Bass.

Gao, L., Liu, X., & Lioliou, E. (2015). A double-edged sword: The impact of institutions and political relations on the international market expansion of Chinese state-owned enterprises. *Journal of Chinese Economic and Business Studies, 13*(2), 105–125.

Haleblian, J., Ji-Yub, K. I. M., & Rajagoplan, N. (2006). The influence of acquisition experience and performance on acquisition behavior: Evidence from the U.S. commercial banking industry. *Academy of Management Journal, 49*(2), 357–370. doi:10.5465/amj.2006.20786083

Haspeslagh, P. C. & Jemison, D. B. (1991). *Managing acquisitions: Creating value through corporate renewal*. Sydney: The Three Press.

Hébert, L., Very, P., & Beamish, P. W. (2005). Expatriation as a Bridge Over Troubled Water: A Knowledge-Based Perspective Applied to Cross-Border Acquisitions. [Article]. *Organization Studies (01708406), 26*(10), 1455–1476. doi:10.1177/0170840605057067

Hitt, M., Harrison, J., & Ireland, R. D. (2001). *Mergers & acquisitions: A guide to creating value for stakeholders*. New York: Oxford University Press.

Javidan, M., Pablo, A. L., Singh, H., Hitt, M., & Jemison, D. (2004). Where we've been and where we're going. In A. L. Pablo & M. Javidan (Eds.), *Mergers and acquisitions: Creating integrative knowledge* (pp. 245–261). Oxford: Blackwell Publishing.

Kale, P. & Singh, H. (2007). Building firm capabilities through learning: The role of the alliance learning process in alliance capability and firm-level alliance success. *Strategic Management Journal, 28*(10), 981–1000.

Kiching, J. (1967). Why do mergers miscarry? *Harvard Business Review, 45*, 84–107.

Kim, J.-Y., Jerayr, H., & Finkelstein, S. (2011). When firms are desperate to grow via acquisition: The effect of growth patterns and acquisition experience on acquisition premiums. *Administrative Science Quarterly, 56*(1), 26–60.

Larsson, R. & Finkelstein, S. (1999). Integrating strategic, organisational, and human resource perspectives on mergers and acquisitions: A case survey of synergy realisation. *Organisation Science, 10*(1), 1–26.

Liu, Y. & Woywode, M. (2013). Light-touch integration of Chinese cross-border M&A: The influences of culture and absorptive capacity. *Thunderbird International Business Review, 55*(4), 469–483. doi:10.1002/tie.21557

London Metal Exchange. (2015). LME copper, 1/3, accessed 1 May 2015, from https://www.lme.com/metals/non-ferrous/copper/

Luo, Y., Xue, Q., & Han, B. (2010). How emerging market governments promote outward FDI: Experience from China. *Journal of World Business, 45*(1), 68.

Marks, M. L. & Mirvis, P. H. (2001). Making mergers and acquisitions work: Strategic and psychological preparation. *Academy of Management Executive, 15*(2), 80–92. doi:10.5465/ame.2001.4614947

Marks, M. L. & Mirvis, P. H. (2011). Merge ahead: A research agenda to increase merger and acquisition success. *Journal of Business & Psychology, 26*(2), 161–168. doi: 10.1007/s10869-011-9219-4

Miles, M. B. & Huberman, A. M. (1994). *Qualitative data analysis: An expanded sourcebook*. California: Sage Publications, Inc.

Ministry of Commerce. (2014). *2013 Statistical bulletin of China's outward foreign direct investment*. Beijing: China Statistics Press.

OZ Minerals. (2008a). ASX release: Updated financial information. Retrieved 1 July 2010, from http://www.ozminerals.com/Media/docs/ASX_Updatedfinancialinformation_081204%20-98110654-c1c2-4679-bf7e-e20fcb9ec048-0.pdf

OZ Minerals. (2008b). Chairman and Managing Director and CEO's speeches and presentations. Retrieved 1 November 2011, from http://www.ozminerals.com/Investor-Information/ASX-Releases.html

Peng, M. W. (2006). *Global strategy*. Cincinnati: Thomson South-Western.

Pennings, J. M., Barkema, H., & Douma, S. (1994). Organizational learning and diversification. *Academy of Management Journal, 37*(3), 608–640. doi: 10.2307/256702

Ralston, D. A., Terpstra-Tong, J., Terpstra, R. H., Wang, X., & Egri, C. (2006). Today's state-owned enterprises of China: Are they the dying dynosaurs or dynamic dynamos? *Strategic Management Journal, 27*, 825–843.

Schoenberg, R. (2006). Measuring the performance of corporate acquisitions: An empirical comparison of alternative metrics. *British Journal of Management, 17*(4), 361–370. doi:10.1111/j.1467-8551.2006.00488.x

Scott, W. R. (2014). *Institutions and organisations: Ideas, interests, and identities* (4th ed.). Los Angeles: Sage Publications.

Shimizu, K., Hitt, M. A., Vaidyanath, D., & Pisano, V. (2004). Theoretical foundations of cross-border mergers and acquisitions: A review of current research and recommendations for the future. *Journal of International Management, 10*(3), 307–353. doi:10.1016/j.intman.2004.05.005

Shrivastava, P. (1986). Postmerger integration. *Journal of Business Strategy, 7*(1), 65–76.

State-owned Assets Supervision and Administration Commission (SASAC). (2015). A list of centrally-controlled state-owned enterprises. Retrieved 1 November 2011, from http://www.sasac.gov.cn/n1180/n1226/n2425/index.html

Thornton, P., Ocasio, W., & Lounsbury, M. (2012). *The institutional logics perspective: A new approach to culture, structure, and process*. Oxford: Oxford University Press.

UNCTAD. (2014). World investment report. Retrieved 26 June 2014, from http://unctad.org/en/PublicationsLibrary/wir2014_en.pdf

Weber, Y. & Shenkar, O. (1996). National and corporate cultural fit in mergers/acquisitions: An exploratory study. *Management Science, 42*(8), 1215–1227.

Williamson, P. J. & Raman, A. P. (2011). How China reset its global acquisition agenda. *Harvard Business Review, 89*(4), 109–114.

Yin, R. K. (2003). *Case study research: Design and methods*. Thousand Oaks: Sage Publications.

Zollo, M. & Singh, H. (2004). Deliberate learning in corporate acquisitions: Post-acquisition strategies and integration capability in U.S. bank mergers. *Strategic Management Journal, 25*(13), 1233–1256. doi:10.1002/smj.426

Zollo, M., & Winter, S. G. (2002). Deliberate learning and the evolution of dynamic capabilities. *Organization Science, 13*(3), 339–351.

5
Corporate Governance in Chinese-Controlled Subsidiaries

Corporate control by MNCs of their foreign subsidiaries has become increasingly complex and challenging with a rapid increase in levels of globalisation and advancement of technology, as well as the growing participation of MNCs from emerging economies such as BRICS countries (Brazil, Russia, India, China and South Africa). Consequently, the revenues and profits generated by foreign subsidiaries can represent substantial amounts to the MNCs. Moreover, these foreign subsidies can possess considerable levels of technological and management know-how that are important for the competitive advantages of the MNCs, particularly those from emerging or developing economies and those with subsidiaries in developed countries.

Corporate control has long been an important topic for studies due to its potential major impact, as has been seen since the enactment of the Joint Stock Act in the United Kingdom in 1844 when a new concept of economic entities named "corporations" was introduced. Corporations have made it possible to separate ownership from control in corporate activities, which in turn can give rise to management control in powerful corporations due to the wide range of ownership and thus lack of effective monitoring by the owners. Consequently, management control has become a new way of organising our business activities and competing with traditional thinking or logic of owner control. Berle and Means (1932) warned of the ascendance of such management control and its potential impact on shareholders as well as society. The potential problems caused by management control were theoretically analysed by Williamson in his work on agency theory (Williamson, 1981).

Corporate governance has emerged as a new competing logic in corporate control in developed countries since the late 1970s, although the concept of corporate governance is as old as the corporate form itself (du Plessis et al., 2010). Corporate governance is defined as "the system by which companies are directed and controlled" (Institute of Directors in Southern Africa, 1994) and has been extensively researched by several disciplines, particularly law, finance and economics. Much of the scope of these studies has primarily focused on the agency problems caused by the separation of ownership and control and thus on the examination of management contracts and several major internal and external corporate control mechanisms, such as incentives of managers, the role of the board, ownership concentration and market for corporate control in order to better align management interest with the owners (Renders & Gaeremynck, 2012; Tam, 1999; Young et al., 2008).

Recent studies on corporate governance have shifted the focus from these internal and external control mechanisms to the role of institutions in corporate control and have considered corporate governance as being embedded in institutions. Friedland and Alford (1991) further argued that corporate governance is an institutional logic, or *organising principle*, used by society in structuring our social systems and guiding our practices.

Building on the seminal work of Friedland and Alford (1991) on institutional logics, several researchers (Lounsbury, 2001; Thornton & Ocasio, 1999, 2008) further developed a theoretical framework, "the institutional logics perspective" (ILP), as "a meta-theoretical framework for analysing the inter-relationship among institutions, individuals, and organisations in social system" (Thornton et al., 2012, p. 2). The emergence of ILP as a meta-theory has offered a new framework to examine corporate control from a more comprehensive and systematic perspective. Specifically, it offers a new framework for analysing the characteristics of an intuitional logic, its dimensions (materials practice and cultural symbols) and how it emerges and evolves (Thornton et al., 2012). Nevertheless, it is still under development and thus needs empirical testing and further refinement. Moreover, little research has been done on how institutions influence corporate control, particularly MNCs from transitional economies.

Corporate control in China has been in a transitional stage since 1979 when China implemented its "open-door policy", thus it can offer a golden research opportunity for studying the dynamics of institutional logics and its further development. As Thornton and her colleagues pointed out:

> [T]he current transformation of China offers a natural, real-time laboratory to study the origins, interdependence, stability, and

> change of the different institutional orders of the inter-institutional system.
>
> (Thornton et al., 2012, p. 174)

This chapter focuses on the corporate control practices of Chinese-controlled and publicly listed subsidiaries in Australia and how they are influenced by the corporate control logics at their parent companies. We look at how these subsidiaries are moderated by their resources and at the characteristics of their Chinese outward foreign direct investment (OFDI). This chapter also aims to explore the benefits and limitations of applying the newly developed institutional logics framework to the area of corporate control in Chinese firms. China has become an important source of global capital, ranking the fifth largest in the world for consecutive years since 2009 (UNCTAD, 2014). However, the question of how these emerging Chinese MNCs manage their corporate governance system and control their subsidiaries needs to be further investigated.

Based on this consideration, this chapter is organised according to the following structure. We first describe the concept of institutional logics and its characteristics and then discuss the evolution of corporate control in Chinese firms by examining these characteristics. We then develop our hypotheses, followed by the description of research methods employed in this study. The last two sections present the findings and discuss the benefits and limitations of applying ILP to analysing corporate control in Chinese MNCs.

The institutional logics perspective and the logics of corporate control

The institutional logics perspective

The institutional logics perspective has recently been developed as a meta-theoretical framework for institutional analysis. Lounsbury (2007) referred to the concept of logics as "broader cultural beliefs and rules that structure cognition and guide decision making in a field" (p. 289). The concept of institutional logics was originally presented by Friedland and Alford (1991) and defined as "a set of material practices and symbolic constructions, which constitutes *its organising principles*" (p. 248). It has been further developed by Thornton and Ocasio (1999) as "the socially constructed, historical pattern of material practices, assumptions, values, beliefs, and rules" (p. 804). While Friedland and Alford's (1991) definition emphasised the latent nature of institutional logics, Thornton and Ocasio's (1999) definition seems to emphasise the physical and cultural manifestation of institutional logics, which has greatly facilitated the understanding and analyses of institutional logics, and their measurement.

The institutional logics perspective views our society as an inter-institutional system of institutional orders, such as Christian religion, the nuclear family, the bureaucratic state, the capitalist market and democracy (Friedland & Alford, 1991). Each of these institutional orders has its own principles, practices and symbols. These principles, practices and symbols shape the context in which reasoning takes place and rationality is perceived. In other words, institutional logics represent a framework of references that condition the choices of actors (individuals and organisations) for sense-making, the vocabulary they use to motivate action and their sense of self and identify under different institutional orders. These actors, or "agency" in Scott's (2008) terms, are aware of these differences among institutional orders and incorporate them into their decision-making, thoughts and beliefs. At the same time, the actors' behaviours and vocabularies can shape changes in the institutional logics.

Proposed as a meta-theory, the institutional logics perspective is principally characterised by (1) the duality of agency and structure; (2) institutions as material and symbolic; (3) institutions at multiple-levels of analysis; and (4) institutions as historically contingent (Thornton & Ocasio, 2008; Thornton et al., 2012).

Institutional logics have been applied to studies in a number of areas and at different levels, such as organisation theory, management, cultural sociology and cognitive psychology. These studies include the shift from editorial to market in the higher education publishing industry (Thornton & Ocasio, 1999), practice diffusion from a regulatory logic to a market logic in mutual fund industry (Lounsbury, 2002). They have also been used to analyse contemporary phenomena such as occupation prestige ranking (Zhou, 2005), corporate social responsibility, practice adoption in international financial reporting standards, consumer behaviour at market level, stability and change in the public employment service sector, and corporate ethical decision-making and behaviour in the context of MNCs (Kostova et al., 2008; Tan & Wang, 2011). However, few studies have been conducted in transitional economies or in the area of corporate governance and control.

Corporate control in China: The evolution of institutional logics, social values and corporate practices

Corporate control in Chinese firms presents a very unique situation and has been in transition since 1949 when the Chinese Communist Party

(CCP) took over. Since then, the logics of corporate control in China have undergone three stages of changes from state/bureaucrat control to bureaucrat/manager control and then to a hybrid system of bureaucrat/manager and modern corporate governance. The following sections examine the change of structure, practice and symbolic representation of corporate control in these three stages using the institutional logics perspective.

Stage I: Dominant state control (1949–1979)

Soon after the CCP took over in 1949, almost all business entities in China were confiscated and owned by the government at different levels. Since then, the government has been the owner of business organisations and has been extensively and directly involved in their operations. Strictly speaking, all Chinese commercial entities became corporatised as they were owned by numerous shareholders and transformed into Chinese public ownership system.

Politically, the state control logic over national business activities is an important control instrument of socialist ideology. Socially, most Chinese people, including contemporary political and business leaders, have grown up and been educated with such a political ideology. As a result, the logic of state control over economic activities *at the national level* has been established and deeply institutionalised both as material practices and cultural symbol in terms of economic structure and coordination systems at the societal level.

Government agencies were structured accordingly around this state control logic, aimed at implementing these practices, as were the development of regulatory, legal and administrative systems across government and organisation levels. More specifically, the government agencies conducted many micro-economic activities, in addition to exercising macro-economic control. These activities included business planning, pricing, allocation of production and distribution. Probably the most important intervention used by the government was the appointment of the Party Secretary and senior managers to companies. Business organisations were treated as cost centres or working units (the so-called *danwei*). The underlining institutional logic was that the state was responsible for society's employment, welfare and production and allocation of goods and services.

Economic structures, national business systems (such as banking, administrative systems, accounting and tax) and processes were established and organised around this dominant institutional logic. The deeply institutionalised state control logic, coupled with the appointment of senior managers to the business organisation, resulted in a bureaucrat/manager-control logic at the organisational level with bureaucrats being heavily dominant as they acted as proxies of state ownership and commanded most activities, commercial or non-commercial (e.g. employees housing, social security and health care), within the organisation.

At the organisational level, Chinese firms were organised as production plants with the plant head responsible for functional operations, as well as having other non-commercial duties. Managers were appointed by the government; production inputs and outputs were planned by the government. Employment unions were set up primarily for looking after the employees' welfare and Chinese firms were responsible for providing free accommodation to their employees. In summary, a business enterprise was set up and operated like a mini-government. Thus, the bureaucrat/manager-control logic prevailed in Chinese firms before 1979.

Stage II: A hybrid of bureaucrat/manager control (1979–1992)

The "open-door policy" and economic reform implemented since 1979 have gradually delegated responsibilities from bureaucrats or government agencies to state-owned enterprise (SOEs) managers. Reforms included the introduction of performance incentives, price liberalisation and profit retention, as well as contracting responsibilities to managers (Tenev & Zhang, 2002). This gradually reduced the degree of government intervention in the operations of business organisations. However, no fundamental changes were made in the ownership of SOEs, although China started its experiments on the ownership reform in 1984 with 11 SOEs becoming shareholding enterprises. The most important change in corporate control was probably in the government's *practices* as reflected by the reduction of government intervention and the strengthening of management control in SOEs. Consequently, there was a shift in corporate control to a more balanced bureaucrat/manager control. There was no fundamental change in the *structure* of corporate control in China during this period.

Stage III: The installation of corporate governance as a competing logic with bureaucrat/manager control (1993–present)

Substantial changes have been witnessed over the last two decades in corporate control in China, particularly in its structure, ownership and symbolic representation.

From a symbolic perspective, the logic of corporate governance emerged as a new institutional logic for corporate control in China in 1993 when the Third Plenary Session of the 14th Party Congress issued the Decision on Issues Concerning the Establishment of a Socialist Market Economic Structure. Subsequently, China formally introduced the concept of corporation through importing (Roche, 2006, p. 36) as a crucial part of the China's effort to create a "modern enterprise system" and diversify its state ownership. This started a new wave of enterprise reform in China focusing on the corporatisation and ownership diversification of SOEs.

As part of the market-oriented reforms, corporatisation and ownership diversification saw many large and medium SOEs being corporatised and almost all small SOEs being bought out by their managers, employees or individuals, often dominated by inside directors. As a result, many larger SOEs have been listed, often as part of the corporation, on domestic or international stock exchanges.

With regard to company structure, the Company Law was promulgated in November 1993, introducing the concept of a board of directors for large corporations, and stipulating the role of auditing. China had already established two stock exchanges in 1990 and 1991, as well as the China Securities Regulatory Commission (CSRC) in October 1992. International and domestic institutional investors have been gradually encouraged to invest in Chinese stock exchanges. Figure 5.1 outlines the key actors and structure of corporate governance in China.

Although the Chinese government has imported the concept of corporate governance, restructured its administrative systems and structures, promulgated corporate laws and gradually delegated many responsibilities to business organisations, the shift of corporate control logic from bureaucrat/manager controls to corporate governance has proven to be very difficult for several reasons. Culturally, the concept of corporate governance has been implemented in practice, but not internalised with respect to its value and meaning among the key Chinese actors, such as government agencies, business executives, board directors, institutional investors and, more importantly, the general public. The implementation of corporate governance can have a negative impact on the power and interest of Chinese bureaucrats and managers in the context of bureaucrat/manager control logic.

In practice, the Chinese government is still the owner or controls a significant proportion of organisations' shares, particularly large Chinese organisations, and has still been heavily involved in business activities at the organisational level, particularly in SEOs. This approach includes

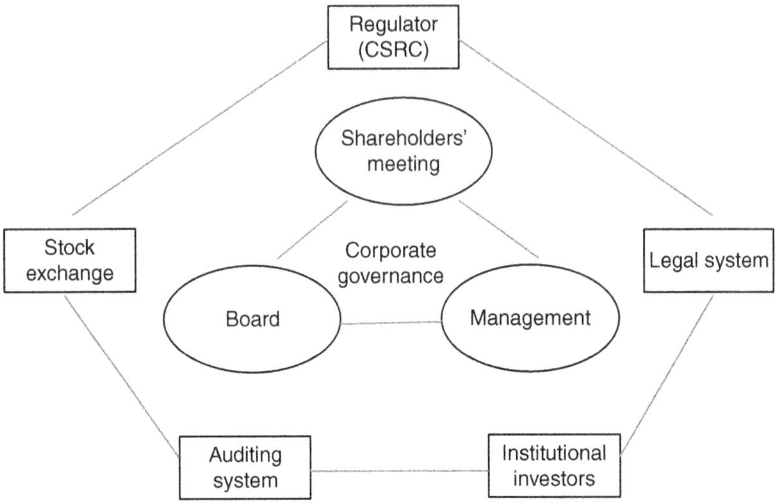

Figure 5.1 The relationship of institutional players within the context of corporate governance in China
Source: Kang et al. (2008).

appointing senior managers to SOEs, coordinating inter-firm activities and financing.

Structurally, the Chinese government has installed key institutions and established relevant organisations for implementing corporate governance. Nevertheless, the devil is in the details in terms of practicing corporate control.

The above discussion shows that the implementation of corporate governance in China has been unbalanced from the institutional logics perspective, with substantial changes in structures, but less significant changes in practices and symbols (value and meaning). Corporate governance in Chinese corporations can still be described as a control-based model (Liu, 2006) with major control exercised by the principal controlling shareholder, often the Chinese government. As Tenev and Zhang (2002) concluded in their studies of corporate governance in China, "ownership diversification and corporatization have had only a limited impact on corporate behaviour".

The above illustration not only demonstrates the importance of internalisation of a new institutional logic in society, but also shows that a balanced approach is needed when introducing a new institutional logic, particularly when it competes with the existing logic. Figure 5.2

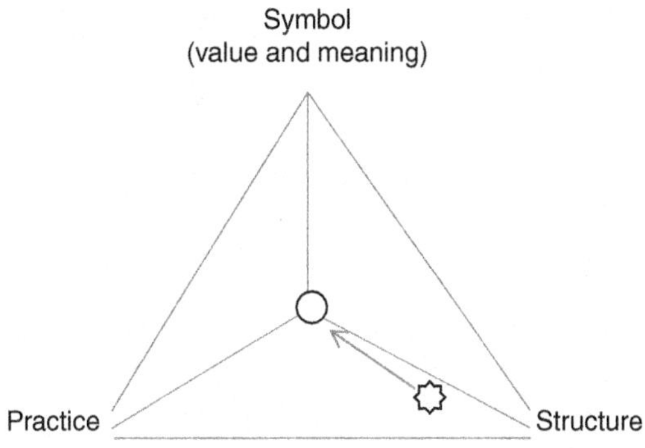

Figure 5.2 The approach adopted by the Chinese government in implementing corporate governance

outlines the current approach of implementing corporate governance in China and indicates the potential future direction.

Following the definition of institutional logics (Thornton et al., 2012) as historical patterns of cultural symbols and material practices, as well as the approach used by Thornton (2004) in treating the institutional orders and its categories as X-axis and Y-axis respectively, the institutional logics in China of bureaucrat/manager control, corporate governance and its characteristics, can be outlined in Table 5.1.

The logics of state control and corporate governance have been competing in China since 1993 and have created "institutional complexity" (Greenwood et al., 2010) for both business executives and bureaucrats. This not only creates inter-institutional contradictions across at the level of state and organisation, but also creates intra-institutional contradictions at each of these levels, since corporate governance, as a new institutional logic of corporate control, demands some fundamental changes in practices, structure, systems and values at both levels.

Hypothesis development

Corporate governance practice and its adoption in Chinese-controlled foreign subsidiaries

As discussed above, corporate governance as a major mechanism of corporate control was introduced in 1993 and could be described as imported and installed as an essential part of establishing "modern enterprise systems" and primarily for raising money for the Chinese government (Liu, 2006). Thus, a unique model of corporate governance

Table 5.1 Institutional logics of bureaucrat/manager control and corporate governance

Organising principles/ categories	Bureaucrat/manager control	Corporate governance
Cultural symbols (meaning and ideation)		
The role of the state	State plans the inputs and outputs, allocates resources to organisations and appoints senior managers with administrative control and central planning	Managers are appointed by: (1) the CCP (2) parent company (3) the board
The role of the corporation	Cost centres with multiple objectives A political institution, an administrative body and an economic producer No economic incentive, no clearly defined boundaries	Conducting national economic activities in SOEs and controlling listed companies Wealth maximisation for shareholders, particularly the major shareholder (i.e. the state and other institutional owners)
Source of legitimacy and authority	Providing employment and social welfare Run by bureaucrats	Market position of the firm Top management leadership

Material practices and structure		
Informal control mechanism	Industry networks	Organisation culture
Employment	Life-time (permanent) employment	Mix of life-time employment and fixed-term contract
The role of managers	Proxy of the state Responsible for all activities Head of the plant with technocrat capability	Proxy of the state or corporate professional managers
Structure	Primarily conglomerate ("small government") and functional Inter-firm coordination by industry-specific ministry/bureau at national level/provincial or city levels	Corporations as independent economic entities Industry-specific ministry replaced by industry association
System	Business systems coordinated by state	Business system coordinated by the corporation
Monitoring	By state agent without effective ultimate principal	By board

has been adopted in China. This model is control-based and characterised by (1) the controlling shareholders (often the state) who employ all feasible governance mechanisms to tightly control the listed firms; and (2) a management-friendly board (Liu, 2006). More specifically, "concentrated ownership structure, management-friendly boards, inadequate financial disclosure and inactive take-over markets have been the standard governance practice commonly observed among the Chinese listed firms" (p. 418).

The existence of a single dominant shareholder is probably the most distinctive characteristic of corporate governance in China. Liu (2006) conducted a survey on the listed companies in China and found that the largest shareholder in China on average held 44.8 per cent of shares, while the sum of the percentage of share holding held by the second to the tenth largest shareholders held an average of only 16.93 per cent. Moreover, a majority (55.6 per cent) of the companies were controlled by the government and their shares were not tradable. Thus, there is no take-over market. More than two-thirds (78.9 per cent) of the publicly listed firms in China have a parent company that can render the listed company's operations more complex and also reduce its transparency. An overwhelming majority of these parent companies are SOEs, with senior managers often appointed by the Party from government bureaucrats, producing a dominant bureaucrat/manager-control logic in these enterprises. This ownership concentration, coupled with the existence of parent companies, has resulted in the selection of most senior managers in listed companies from their parent company.

Regarding the selection and composition of the board of directors, Chen et al. (2004) reported that almost 50 per cent of the directors are appointed by state-controlled owners, and another 30 per cent are affiliated with various layers of government agencies. There are few professionals (lawyers, accountants, finance experts) on Chinese boards and almost no representation by minority shareholders. More than one-third (34.6 per cent) of the CEOs are also the chairmen of the board of directors. Additionally, top managers typically own little of their companies' shares (only 0.1 per cent). Therefore, it can be expected that most board of directors are inside or related to the company, and the agency cost is very high in listed Chinese companies as managers hold few shares in the company.

The problem of weak legal and regulatory enforcement in China has also slowed the shift from bureaucrat/manager control to corporate governance and made corporate governance less effective. Although corporate law was introduced in China in 1993, forming the legal foundation of corporate governance, evidence so far shows that it has been ceremonially, rather than substantially, implemented. The adoption of *Corporate Governance Practices* required both implementation and internalisation (Kostova et al., 2008). The above discussion indicates that corporate governance has not been internalised within Chinese political, business and legal communities, although the central government has exerted pressure for it to be implemented since 1993.

Weak corporate governance practice has been identified as the major cause of recent corporate scandals and poor corporate performance; these include corruption, stock market manipulation, tax cheating, fraudulent dealings, plundering of state assets and the lack of protection for minor shareholders rights (Tam, 2002). Consequently, this has generated a perception among Chinese business executives, politicians and the public that corporate governance in China is a very weak corporate control mechanism.

In summary, the legacy of the long-established and practised state control at the national level, and its resulting bureaucrat/manager control at the organisational level, is still lingering and often dominants in many areas of the Chinese economic system. Corporate governance, a fundamental mechanism of organisation control in Western countries, has been ceremonially implemented in Chinese firms due to the difficulty in the internalisation of such a control mechanism. China urgently needs further reform of deeply rooted institutional systems and structures at the national level, as well as transforming corporate control practices by following modern corporate governance practices at firm level.

From an institutional logics perspective, we expect that the corporate control practices adopted by the Chinese MNCs are likely to be applied to their investments overseas for several reasons. First, these practices, as the material element of institutional logics, have long been undertaken in Chinese firms. The practices are well served by the organisational structure, systems and decision-making processes within the Chinese organisations and are endorsed by key stakeholders (e.g. government agencies and banks) in China. Second, these practices are legitimised and have been deeply embedded in Chinese business culture and business executives' cognitions as the symbolic element of social-culture and individual cognition. The latter aspects create difficulties in shifting from one logic of corporate control to another. As argued by Fligstein (1990), executives' view on how to best run the corporation are selectively influenced by their professional background and experience in the corporation. Third, there are uncertainties and costs involved in restructuring the organisation (i.e. changing rules, systems and organisational reporting lines). Based on these considerations, we develop the following hypotheses:

> H1. Chinese MNCs prefer appointing inside and related directors to their foreign subsidiaries rather than outside independent and professional directors.
>
> H2. Chinese MNCs prefer using manager control in controlling their foreign subsidiaries rather than corporate governance.

Moderating the relationship between institutional logics and corporate control practices: Organisational resources and size of foreign subsidiaries

Organisational resources

According to the institutional logics framework proposed by Thornton et al. (2012), the resource environment affects the generation of practices, but is also materially shaped by practices through the consumption and production of resources. From the institutional logics perspective, resources can influence organisational practices at two levels: societal and organisational levels. At the societal level, the resource environment, such as that in economic systems (i.e. manpower, finance, physical capital and customer demands), and forms of governance (i.e. market, governments, corporations and information network) provide the opportunities and constraints for conducting material practices. It can be expected that organisations draw such resources from their environment for undertaking practices, such as corporate control. At the organisational level, the resources controlled by the organisation can influence the types and the quality of practices or "sets of meaningful activities" based on the definition of practices (Thornton et al., 2012, p. 128). At the same time, these practices can generate new resources, particularly intangible and culturally embedded, for the organisation and society.

With specific reference to corporate control, on the one hand, the adoption of bureaucrat/management logic has generated a large pool of managers who are capable of managing their businesses or have careers in the Chinese institutional and business environment. On the other hand, the lack of emphasis on corporate governance has not produced any real demand for professional directors in the corporate boardroom. Most board of directors in Chinese firms are ceremonially recruited (Liu,

2006), either from the parent company or from government agencies, and this has been regarded as a rubber stamp for corporate decisions.

Therefore, many Chinese firms have a shortage of directors who are knowledgeable and experienced in corporate governance (Hu & Tam, 2012), particularly in the context of foreign subsidiary operations. However, some large Chinese MNCs may be exceptions as they were set up for conducting international business (Cai, 1999) and thus have gained much experience and accumulated a talent pool from their international business network and operations.

Size and complexity of the foreign subsidiary's operations

Another factor moderating the relationship between the institutional logics of corporate control in home country and its corporate control practices in a foreign subsidiary is the size and complexity of the foreign subsidiary's operations. The role played by a foreign subsidiary can vary substantially and have different impacts on the parent corporation in terms of revenue, profits and corporate strategy. In the mining industry, the operation complexity increases as a mining company develops from the stage of exploration, to construction and production. Exploration is relatively simple in terms of the scope of operations, capital spending, workforce and risk management and value network management (suppliers, transportation, traders and buyers), compared with those at the stages of construction and production. Size is often indicated by the amount of investment, particularly investment in operations. To some extent, the size and complexity of a foreign subsidiary are a resource-related issue; the larger the size and the more complex the operation, the more demands made on corporate governance and management in a foreign subsidiary. Therefore, we develop the following hypotheses:

> H3a. The appointment of inside directors to Chinese subsidiaries is moderated by the Chinese MNCs' internal HR resources.
>
> H3b. The appointment of inside senior managers to Chinese subsidiaries is moderated by the Chinese MNCs' HR resources.
>
> H4. Chinese MNCs are more likely to appoint non-Chinese senior managers to their large foreign subsidiaries with complex operations.

The relationships among the logics of corporate control in China, corporate practices in foreign subsidiaries, organisational resources and the size and complexity of foreign operations are outlined in Figure 5.3.

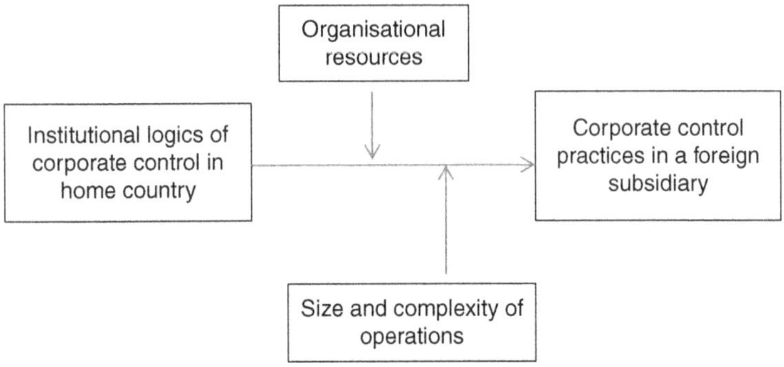

Figure 5.3 Relationships between the logics of corporate control in China, corporate practices in foreign subsidiaries, organisational resources and the size and complexity of foreign operations

Research methods

This chapter is exploratory in nature and part of a larger research project on Chinese investment in Australia. Based on a database developed by one of the authors and his research team, we have used the following three criteria in selecting the sample: Chinese investment of more than A$10 million; Chinese share in their foreign entity more than 40 per cent (an effective control majority), but below 100 per cent; and public listing (either the subsidiaries or the partner in the case of a joint venture [JV]). Ten Chinese subsidiaries operating in the Australian mining industry met these three criteria and were thus selected. These Chinese subsidiaries were all operating in the Australian mining industry at the time of data collection. Nine of these companies, including two parent companies of their Australian partners in the case of JVs, are publicly listed on the ASX and one on the Hong Kong Stock Exchange. Therefore, detailed information about the board of directors and the appointment of the senior management team was available from their websites.

The websites of both parent companies were also examined to identify the composition of the boards of directors and senior management team. Five senior managers from four Chinese subsidiaries were then interviewed as part of a larger scale project that explored issues of strategic management in Chinese subsidiaries. These interviews lasted from one to two-and-a-half hours. Table 5.2 presents a summary profile of the ten Chinese subsidiaries, including their parent company, major operations and extent of the headquarters' investment.

Table 5.2 Profile of the sample of Chinese subsidiaries and the interviewees

Chinese FDIs	Description of the parent company	Major operations	Chinese ownership (%)	Operation, stage and extent of investment
S1	A large provincial SOE in gold mining industry	A very large gold mining company in China	51	Gold mining, production (A$225 million)
S2	A very large SOE with a long history in coal mining	Coal mining	72	Coal mining, production (A$3.4 billion)
S3	A very large SOE with more than 30 years of international trade	Conglomerate (vertical integration in non-ferrous metals, mining, processing and trading)	71.8	Base metals mining, production (US$1.3 billion)
S4	The largest gold mining company in China	Chinese CEO, other five senior managers are non-Chinese	81.5	Gold mining, production (A$229 million)
S5	A large private steel company	The largest private integrated steel company in China	46	Iron production and exploration (more than A$30 million)
S6	Conglomerate (vertical integration in nuclear power)	Conglomerate (vertical integration in nuclear power)	70	Uranium mining, exploration (A$81.4 million)
S7	A large Chinese privately owned enterprise	Privately owned enterprise, conglomerate	55	Base metals, development (US$140 million)
S8	A provincial Chinese conglomerate, listed in Shenzhen Stock Exchange, 37% owned by the provincial SOE	Conglomerate (vertical integration in non-ferrous metals)	50.1	Base metals, development (A$45.5 million)
S9	A very large Chinese steel mill	A very large integrated steel mill (mining, processing and trading)	60 (JV)	Iron ore mining, exploration (A$186 million)
S10	A very large Chinese steel mill	A very large integrated steel mill (mining, processing and trading)	50 (JV)	Iron ore mining, development (A$1.4 billion)

Findings and discussion

The following sections report on company ownership and board composition, as well as the appointment of the senior management team and the members' qualifications. Additionally, other control mechanisms used by Chinese MNCs are briefly reported and discussed.

Board of director and chairperson appointments in Chinese MNC subsidiaries

Table 5.3 shows the board of directors and the experience/qualifications of its members at ten Chinese subsidiaries or JVs in the Australian mining industry. As Table 5.3 reveals, the proportion of the number of Chinese directors is closely related to the percentage of Chinese ownership in the subsidiary. It is important to point out that almost all (32/33) of the Chinese directors appointed are inside directors. Overall, this finding supports H1. Nevertheless, three (S1, S7 and S10) of the ten entities have appointed non-Chinese board chairmen and four (S1, S3, S5 and S7) of them have not appointed a majority of their directors to the board. These contradictions need further examination.

S1 and S7 have both appointed non-Chinese chairmen to the board and less than half of the directors to their Australian entities where they hold a majority share. S7 is a large Chinese private conglomerate. It has little experience of international operations and the mining industry. S1 was taken over by a Chinese company a short time before the data collection. Negotiating how many directors are appointed by the Chinese parent company could be part of the investment negotiation process. Given this is a related investment, it could be expected that the Chinese parent company would like to exercise its control over the entity. Thus it would be interesting to observe the corporate control mechanisms used over time.

S10 is a JV with ownership equally divided between a large Chinese steel company and a listed Australian mining company. The Chinese company came to Australia to invest in 2006 and set up the JV, engaging in activities from exploration to development and production. It also holds about 37 per cent of the parent company of its Australian partner and appoints three directors to its nine-seat board of directors. The chairman of the Australian partner company is also the chairman of the JV as it accounted for more than 95 per cent of assets of the Australian parent company. It is evident that Chinese control has been exercised through the board of the Australian parent company, rather than at the

Table 5.3 Board of directors and the experience/qualifications of its members at ten Chinese subsidiaries or JVs in Australia

	Chinese share (%)	Listed	Chinese director(s)/ No. of directors	Chair	Insider directors?	Backgrounds	Other non-Chinese directors
S1	51	A	3 (NED)/7	A, NED C, deputy chair, NED	Yes	Management, finance, geology	Accounting, finance, mining and geology
S2	72	A	6 (all NED)/11	C, NED C*	Yes	Engineering (4), accounting (2)	Five NEDs (accounting (2), transportation & OHS, investment, diplomatic)
S3	72	HK	4 (all NED)/9	C, NED	Yes	International trade/economics (2), finance, management	Three NED (metallurgical engineering, accounting, law, one from HK), Two ED (CEO (chemical engineering), CFO (finance)
S4	89	A	3 (2 NED)/4	C, NED, C*	2 Yes	Geology, engineering (2)	One NED from HK (entrepreneurship)
S5	45	A	2 (all NED)/6	C, NED	Yes	International trade, engineering	Accounting, economics, law, finance
S6	66	A	5 (4 ED)/7	C, NED	Yes – 4 of 5, 1 related	Exploration, financial; geology engineering (2)	Two OZ's NED (one former MD geologist, one with engineering background)

Table 5.3 (Continued)

	Chinese share (%)	Listed	Chinese director(s)/ No. of directors	Chair	Insider directors?	Backgrounds	Other non-Chinese directors
S7	55	A	2 (all NED)/6	A, NED	Yes	Entrepreneurship; finance, investment	Four OZ's NEDs (commerce and investment, law, geological survey and investment)
S8	50.1	In Ain SZ	3 (all NED)/6	C, NED, C*	Yes	Management; commerce; mining engineering	1 OZ's ED (MD, engineering); 2 OZ's NED (investment, law)
S9	60	JV, No	3 (all ED) /5	C, ED	Yes	Mining, finance, mineral processing	Two OZ"s NED (mining engineering, commerce)
S10	50	JV, No	2 (all NED)/4	A, NED	Yes	Finance; engineering (both are VPs of the parent company)	The chairman and MD of the OZ parent Co (accounting, commerce)
	Average 61%		33 (7 ED)/65	7(C)/10	32(I)/33	Engineering (13), finance/accounting (8), management/ commerce (8), 4 geology/exploration (4)	Engineering (7), accounting/finance (8), geology (3), management (8), law (4), others (2)

Note: A: Australian; C: Chinese C*: Chair/president of parent company; ED: Executive Director; HK: Hong Kong; JV: joint venture; MD: managing director; NEC: non-executive chairman; NED: non-executive director; SZ: Shen Zhen Securities Exchange; VP: vice president; OHS – organisational health and safety.

JV level. This raises a question regarding the role of the board at the JV level, which may act mainly as a formality, rather than a major control mechanism at the JV level.

Regarding the proportion of directors appointed by the Chinese parent company, four Chinese subsidiaries have appointed less than 50 per cent of the directors to their Australian subsidiaries. In the case of S5, the Chinese parent company only holds 46 per cent of the Australian entity's share. Consequently, it has appointed two directors to the five-seat board of the entity. This is proportional to the percentage of shares held. In the cases of S1 and S7, both might be subject to a transaction agreement and lack of internal resources, as discussed previously.

The case of S3 is very interesting as it presents an extraordinary case for studying the shift of corporate control in Chinese MNCs. The parent company is one of the top 500 ranked by *Forbes* magazine and was set up by the Chinese government in 1950 as its international minerals trading company. S3 was set up as the result of a friendly Chinese acquisition of the majority of assets of a listed Australian mining company. The Australian management team was kept after the transaction, including its CEO. The company has been relisted on the Hong Kong Stock Exchange since 2009. Although the Chinese parent company held 72 per cent of the relisted company, the proportion of directors appointed by the parent company has been reduced from five out of seven in 2010 to four out nine in 2012. However, one of the independent non-executive directors was appointed from Hong Kong.

Table 5.3 also shows the background/qualifications of Chinese directors who are mainly active in the areas of engineering and geology (17/33), compared with 10/32 of non-Chinese directors active in these areas. Accounting and management are two other areas of expertise Chinese which directors can offer. However, all directors with legal background are non-Chinese. This indicates the lack of international legal expertise in Chinese MNCs.

The above discussion has largely supported our H3a, which states that the appointment of directors to Chinese subsidiaries is moderated by the Chinese MNCs' human resource (HR) capabilities. Chinese MNCs are more likely to appoint directors to their foreign subsidiaries based on their HR capabilities, and they recruit non-Chinese directors in the areas where they lack expertise, particularly in "soft" areas, such as law, environmental protection, government relations, community engagement and organisational safety.

Appointment of MD/CEO and senior managers in Chinese MNC subsidiaries

Only three of the CEOs or MDs in the ten Chinese-controlled subsidiaries were Chinese (see Table 5.4) at the time of the data collection. This appears to contradict our hypothesis (H2). Two factors have contributed to this: Australian government regulations and the lack of HR capabilities in Chinese MNCs. First, in approving the Chinese acquisitions, the Australian government requested that the CEOs of two Chinese controlled entities (S2 and S3) "reside primarily in Australia", implying that they should be Australian residents. Second, the Chinese MNCs lacked HR capabilities, either due to their limited human resources and high demand for qualified experts because to their subsidiary's large and complex operations (S1 and S10) or due to the combination of both (S5, S7 and S8).

In the three cases of the Chinese CEOs, S6 and S9 are still at the exploration stage. Their operations are thus relatively simple. S4 is operating in the same industry as the parent (gold mining); therefore, the Chinese company is very familiar with the operations of its Australian entity.

Based on the above discussion, we can see that H2, H3b and H4 are partially supported by the evidence as presented in Table 5.4.

The deeply rooted bureaucrat/management control logic is vividly reflected in the control practice of S2 by its parent company. As mentioned above, the parent company of S2 has long been operating in the same industry of coal mining as S2. When the parent company acquired a large Australian coal mining company in 2009, one of the conditions imposed by the Australian government in approving the investment application was "CEO and CFO reside primarily in Australia". As a result, the CEO of S2 was selected from Australian managers. However, the CEO was not selected as the company's director. This represented an exception in the Australian mining industry. Instead, an executive committee was set up and chaired by the deputy chairman (Chinese) of the board. This executive committee was responsible for major operational decisions of S2. In addition, a young Chinese-Australian was appointed as CFO to meet the conditions imposed by the Australian government. Both arrangements demonstrated that the real control of S2 operations was in the hands of the board's deputy chairman who was also the chairman of the executive committee.

Table 5.4 Profile of CEOs and senior managers in Chinese subsidiaries in Australia

	CEO	Other senior managers	Chinese management team	Australian management team	Investment (A$)	Stage at May 2010	Year of investment
S1	A	10 (A)	None (0/10)	Mining, exploration, finance, HR, technical services	225 million	Production	2012, A
S2	A	9(A), 3(C)	Finance, Corporate service, excluded assets	CEO (management), COO, operations (4), HR, logistics, business development	>6.0 billion	Production	2009, 2012, A
S3	A	5 (A)	None	CEO, CFO, COO, business development, corporate services,	1.3 billion	Production	2008, A
S4	C	6 (A)	MD (Mining engineering)	CFO (Asian), GM (mining engineering), business development (accounting), general counsel (law), exploration (Geology), general counsel (aw), technical service (mining engineering).	229 million	Production	2012, A
S5	A	5 (A)	None	MD (commerce), COO (geology), CFO, consulting, GM (mining)	>30 million	Exploration and production	2007, A
S6	C	3 (C)	MD (PhD, Uranium exploration), CFO, Technical directors (Geology)	None	81.4 million	Exploration	2009, A

Table 5.4 (Continued)

	CEO	Other senior managers	Chinese management team	Australian management team	Investment (A$)	Stage at May 2010	Year of investment
S7	A	5 (A)	None	MD, COO, CFO, commercial and marketing, GM (operations)	140 million	Development and early production	2009, A
S8	A	3 (A)	None	MD , CFO, GM (Americas), general counsel	45.5 million	Development and early production	2009 A
S9	C (MD)	4 (C), 4 (A)	MD (Mining), CFO, Senior Geologist (Two),	COO, project development, exploration and approvals, health, safety & environment, community relations	186 Million	Exploration	2010
S10	A	2(A)	No	CEO, approvals, CFO	1.4 billion	Development and early production	2006
	3/10 MD (CEOs) are Chinese	10 (C) out of 59					

Notes: A – Australian; C – Chinese.

Conclusion and implication

This chapter explores the logics of corporate control within Chinese MNCs and its impact on its control practices in their foreign subsidiaries. Applying an institutional logics perspective to corporate control at the Chinese MNC level and in the international context, we investigated corporate control practices in ten Chinese-controlled subsidiaries in the Australian mining industry. The Chinese government has applied pressure on its domestic firms, particularly large companies, to shift their corporate management from bureaucrat/manager control to a corporate governance system. These emerging Chinese MNCs face "institutional complexity" (Greenwood et al., 2010) due to pressure from multiple institutional forces. This chapter also tests the applicability of the newly proposed institutional logics perspective in the circumstances of corporate control within an international context as "relatively few institutional logics studies" (Thornton et al., 2012, p. 174) have been undertaken in this area. Several specific concluding remarks can be drawn from this study.

The first observation is that Chinese MNCs have applied their corporate control logics by the parent company of their foreign subsidiaries with an emphasis on manager control rather than corporate governance. Thus, the appointment of directors to their foreign subsidiaries is a symbol of control, rather than a provision of services to the subsidiaries. An overwhelming proportion of Chinese subsidiary directors appointed by their headquarters are indeed insiders with the equivalent proportion of ownership. Moreover, most of the directors appointed are qualified in the engineering/geology, accounting and finance, and management fields, but they lack experience in international management, legal, social/cultural and language skills. This pattern of director appointment is similar to that seen in China, thus reflecting the deep-rootedness and stickiness of institutional logics, in this case, corporate control.

Second, Chinese MNCs prefer appointing insiders as CEOs/MDs to their subsidiaries, rather than outsiders. This is primarily due to China's dominant logic of bureaucrat/manager control and partly due to a high level of trust among insiders (but less trust towards outsiders) in Chinese culture. Consequently, the strategic responsibilities often rest with the CEO/MD of the subsidiary. This has resulted in the establishment of a special organisational structure in the case of S2 that gives the Chinese director responsible for executing operations, whereas the Australian CEO becomes a figure head due to the requirement of the

Australian Government that the CEO should be an Australian resident/ citizen.

The third issue is that the task complexity at the foreign subsidiaries and the relative level of competence and expertise at the parent company moderate the senior management positions at the subsidiary. As a result, non-Chinese managers with local experience are appointed to top management positions at the subsidiary after acquisition. The lack of experienced executives and middle managers, coupled with potential language issues, has limited the number of expatriates sent from Chinese MNC headquarters to their Australian subsidiaries. Moreover, Chinese MNCs have appointed senior managers with a background in science and engineering and recruited local managers for working in areas that require Australian social, cultural and legal knowledge, such as handling government approvals, community relationship building and legal procedures.

The last, but nevertheless important lesson is that a culturally embedded and well-institutionalised corporate control logic cannot be easily changed in the international context. Chinese MNCs largely extend the corporate control logic from their domestic firms to those in Australia, a logic that relies heavily on ownership and personal control and allows management control in their Australian subsidiaries. This kind of practice has been considered to be counterproductive.

Based on our research findings, a number of implications can also be identified. Theoretically, our findings show that institutional inertia exists and increases the difficulties and costs of institutional change. Moreover, different institutional logics can co-exist and are competing for survival and dominance.

Practically, implementing a new institutional logic needs to consider both the material practices, cultural symbols and the cross-level effects, such as national, organisational, social and individual level. In the case of the Chinese MNC's implementation of corporate governance as new corporate control logic, the change of structural element of the material practices seems to be easily achieved among the Chinese MNCs. However, it would require more effort and resources to change cultural symbols generated by cultural value and individual cognition. From a culture perspective, if the new logic, in this case, corporate governance, cannot demonstrate that it can offer value to the corporate sector, community and society, it will not be valued by the corporate executives and bureaucrats. From the cognition perspective, this requires training and education of both business managers and government officials.

The findings also offer new evidence of the mutual relationship between the two dimensions of institutional logics: material practices and cultural symbols. The cultural symbols (i.e. social value and individual cognition) shape, and are shaped by, the practice of corporate control. Thus, our research findings offer new evidence that support the theoretical framework of institutional logic.

Another implication from the findings is that there are two competing strategies for Chinese MNCs to move towards their corporate control logics: global vs. multi-domestic. The selection of orienting strategy depends on the international context in which an MNC operates. Where an MNC has developed its expertise in international operations, it can use a global corporate control logic, such as corporate governance, in its subsidiaries globally. However, MNCs from emerging economies may use a multiple domestic strategy during the early stage of international operations, as these companies do not have sufficient international business expertise, and the control logics in their home country are often different from those adopted in other countries where their subsidiaries have been operating. These certainly add further difficulties for emerging MNCs from developing countries in general, and the Chinese MNCs in particular, to fully develop their international business capabilities in a relatively short period.

References

Berle, A. A. & Means, G. C. (1932). *The modern corporation and private property*. New York: Harcourt, Brace & World.

Cai, K. G. (1999). Outward foreign direct investment: A novel dimension of China's integration into the regional and global economy. *The China Quarterly, 160*, 856–880.

Chen, D., Fan, J., & Wong, T. J. (2004). *Political connected CEOs, corporate governance and post-IPO performance of China's partially privatized firms*. Hong Kong: Chinese University of Hong Kong.

du Plessis, J. J., Hargovan, A., & Bagaric, M. (2010). *Principles of contemporary corporate governance* (2nd ed.). Melbourne: Cambridge University Press.

Fligstein, N. (1990). *The transformation of corporate control*. Cambridge, MA: Harvard University Press.

Friedland, R. & Alford, R. R. (1991). Bring society back in: Symbols, practices and institutional contradiction. In W. W. Powell & P. J. DiMaggio (Eds.), *The new institutionalism in organizational analysis* (pp. 232–263). Chicago: The University of Chicago Press.

Greenwood, R., Díaz, A. M., Li, S. X., & Lorente, J. C. (2010). The multiplicity of institutional logics and the heterogeneity of organizational responses. *Organization Science, 21*(2), 521–539.

Hu, H. W. & Tam, O. (2012). Independent directors in China and India: A comparative assessment. *Company and Securities Law Journal, 30*, 453–470.
Institute of Directors in Southern Africa. (1994). King report on corporate governance. Retrieved 1 October 2014.
Kang, Y., Shi, L., & Brown, E. D. (2008). Chinese corporate governance: History and institutional framework. Retrieved 1 February 2013, from www.rand.org
Kostova, T., Roth, K., & Dacin, M. T. (2008). Institutional theory in the study of multinational corporations: A critique and new directions. *Academy of Management. The Academy of Management Review, 33*(4), 994–1006.
Liu, Q. (2006). Corporate governance in China: Current practices, economic effects and institutional determinants. *CESifo Economic Studies, 52*(2), 415–453.
Lounsbury, M. (2001). Institutional sources of practice variation: Staffing college and university recycling programs. *Administrative Science Quarterly, 46*(1), 29–56.
Lounsbury, M. (2002). Institutional transformation and status mobility: The professionalization of the field of finance. *Academy of Management Journal, 45*(1), 255–266.
Lounsbury, M. (2007). A tale of two cities: Competing logics and practice variation in the professionalizing of mutual funds. *Academy of Management Journal, 50*(2), 289–307.
Renders, A. & Gaeremynck, A. (2012). Corporate governance, principal-principal agency conflicts, and firm value in European listed companies. *Corporate Governance: An International Review, 20*(2), 125–143.
Roche, J. (2006). *Corporate governance in Asia*. New York: Routledge.
Scott, W. R. (2008). *Institutions and organisations* (3rd ed.). Los Angeles: SAGE Publications.
Tam, O. K. (1999). *The development of corporate governance in China*. Northampton, MA: Edward Elgar
Tam, O. K. (2002). Ethical issues in the evolution of corporate governance in China. *Journal of Business Ethics, 37*(3), 303–320.
Tan, J. & Wang, L. (2011). MNC strategic responses to ethical pressure: An institutional logic perspective. *Journal of Business Ethics, 98*(3), 373–390.
Tenev, S. & Zhang, C. (2002). *Corporate governance and enterprise reform in China: Building the institutions of modern markets*. Washington, DC: World Bank and International Finance Corporation.
Thornton, P. & Ocasio, W. (1999). Institutional logics and the historical contingency of power in organizations: Executive succession in the higher education publishing industry, 1958–1990. *The American Journal of Sociology, 105*(3), 801–843.
Thornton, P. (2004). *Markets from culture: Institutional logics and organizational decisions in higher education publishing*. Stanford CA: Stanford University Press.
Thornton, P. & Ocasio, W. (2008). Institutional logics. In R. Greenwood, C. Oliver, K. Sahlin, & R. Suddaby (Eds.), *The SAGE handbook of organizational institutionalism* (pp. 99–129). London: Sage Publications.
Thornton, P. Ocasio, W., & Lounsbury, M. (2012). *The institutional logics perspective: A new approach to culture, structure, and process*. Oxford: Oxford University Press.

UNCTAD. (2014). World investment report. Retrieved 26 June 2014, from http://unctad.org/en/PublicationsLibrary/wir2014_en.pdf

Williamson, O. (1981). The economics of organization: The transaction cost approach. *The American Journal of Sociology, 87*(3), 548–577.

Young, M. N., Peng, M. W., Ahlstrom, D., Bruton, G. D., & Jiang, Y. (2008). Corporate governance in emerging economies: A review of the principal–principal perspective. *Journal of Management Studies, 45*(1), 196–220.

Zhou, X. (2005). The institutional logic of occupational prestige ranking: Reconceptualization and reanalyses1. *The American Journal of Sociology, 111*(1), 90140.

6
Cross-Cultural Management and HRM

Introduction

Many Chinese companies operating their business in overseas, namely Chinese MNCs, face the challenges of cross-cultural management. Adequate "hard" and "soft" HRM policies and practices could enable them to overcome these problems. Using case studies of Chinese companies operating in Australia, this chapter focuses on identifying problems, developing adequate policies and implementing them in a culturally appropriate manner. Companies from a transitional economy, such as China, operating in a developed economy, such as Australia, confront new challenges, particularly in the areas of new labour market environment and relevant labour laws and regulations. It is crucial for these "foreign" companies to be familiar with such new institutional and economic environments and be able to adopt and adapt new ways of doing business and managing their workforce. The outcome of the case study analysis could be useful for other companies to learn both positive and negative lessons, and this is particularly relevant to those companies from transition economies operating in developed economies.

A number of challenges confront Chinese companies operating overseas in general and in Australia in particular. Most issues that have been identified relate to the capacity of Chinese MNCs managing their overseas business operation, such as the overall quality of Chinese expatriate managers (i.e. their cross-cultural communications and management capability), enterprise cultures (i.e. adopting cultural elements of both home and host countries and combining them adequately) and the companies' capability to be familiar with and fit into the local industry norms and regulatory environment well. In addition, at the individual level, some cultural and behavioural differences between Chinese managers and local employees could jeopardise possible good working relationships. For instance, Chinese people pay more attention to *guanxi* that emphasises personal relationship with reciprocity and mutual obligation, but Australian people would be more interested in working with each other based on formal contractual obligation, rule and regulation. Face saving behaviour of Chinese managers may also have a negative consequence as managers may fail to address employee problems explicitly, leading to the accumulation of further problems and generation of more confrontations at a later stage. Therefore, it is our intention to present a particular chapter on cross-cultural management and HRM regarding Chinese MNCs operating in Australia. This chapter is designed in the following structure: we review the relevant literature on Chinese MNCs' international HRM (IHRM) policies and strategies in general and their operation in Australia in particular, so we can identify some common themes and patterns. Then we discuss the methodology and a number of key issues are generated. By following these key issues, we present our own case studies with more detailed records of Chinese managers' views. Finally, we discuss and conclude the chapter by summarising the key issues and implications for future research and policy improvement.

Literature review on Chinese MNCs' IHRM policies and strategies

Following the national policy on "going global" in 2004, more Chinese MNCs have been engaging in cross-country investment activities. Following this phenomenon, a number of research projects have gradually appeared in international journals and books with particular attention on the importance of linking IHRM to the realisation of internationalisation strategy and adopting strategic IHRM with different orientations, such as global integration, local responsiveness and/or a combination of both, as well as detailed IHRM functions being operated in a different national context (Cooke, 2008, 2012; Cooke & Lin 2012; Fan et al., 2013; Fan et al., 2012; Luo & Tung, 2007; Shen, 2005, 2006; Shen & Edwards, 2004; Thite et al., 2012; Wang et al., 2013; Warner & Nankervis, 2012; Zhang & Fan, 2014; Zheng, 2013; Zhu et al., 2008).

The relevant literature started the debate about the lack of empirical research that documents the IHRM strategies and practices of international firms that have their headquarters outside North America and whether Western models are equally relevant when applied to MNCs from developing countries like China (Scullion, 1994; Shen,

2004, 2005). Therefore, a number of empirical research projects focusing on IHRM functions were gradually published in the mainstream HRM journals. For example, Shen and Edwards (2004) explored international recruitment and selection policies and practices of Chinese MNCs. They reported that the staffing approaches of these companies could be categorised into three groups: ethnocentric, polycentric and an ethnocentric approach with a strong polycentric tendency. The ethnocentric approach to staffing emphasised parent-country nationals (PCNs), namely home-country expatriates, occupying key positions at subsidiaries. In contrast, the polycentric approach promoted the host-country nationals (HCNs) taking leadership roles in subsidiaries. Hence, the ethnocentric approach with a strong polycentric tendency represented a mixture of coordination, leadership, control and local responsiveness among PCNs and HCNs.

Shen (2006) continued this research, focusing on key factors influencing international staffing in Chinese MNCs. He found that there were some consistencies with existing literature in which host-contextual factors, including political, legal, economic and socio-cultural factors, and firm-specific factors, including international strategy, international organisation structure, organisational culture, stage of internationalisation and type of industry, were major determinates of international staffing policies and practices (Shen, 2006, p. 310). However, the same categories of factors had different details when applied to the Chinese context. For example, political factors were associated not only with political stability but also with party influence, leadership and loyalty. Other findings were that international experience was a catalytic agent of change but not a determinate of international staffing approaches, and economic ownership had a significant impact on domestic HRM, but a negligible impact on IHRM. In addition, size of business, reliance on international markets and top management perception of home HRM systems were determining factors. The evidence also showed that trust and personal moral merits influenced the decision on international staffing (Shen, 2006, p. 311).

Luo and Tung (2007) conducted their investigation based on a springboard perspective to describe the internationalisation of emerging market MNCs, including Chinese MNCs. They claimed that these MNCs use outward investments as a springboard to acquire strategic assets needed to compete more effectively against global rivals and to avoid the institutional and market constraints they face at home. Their "springboard" behaviours are characterised by overcoming their latecomer disadvantage in the global market through proactively acquiring critical resources needed to compete at home and abroad (Luo & Tung, 2007). Chinese MNCs such as TCL, Lenovo, Chunlan, ZTE and Haier reorganised their home supply or production bases to meet their increased global sales for high-end products, or they rebranded their products with acquired foreign technologies and trademarks. Key challenges for these MNCs pertained to post-springboard, post-acquisition difficulties, in particular related to building effective working relationships with host-country stakeholders (including employees), reconciling disparate national- and corporate-level cultures, organising globally dispersed complex activities, and integrating home- and host-country operations (Luo & Tung, 2007, p. 494). However, rapid expansion into new markets often exacerbated the difficulties due to lack of global experience, managerial competence and professional expertise. Challenging issues included how to deal directly and effectively with foreign consumers, regulators, legislators, courts, unions, employees and financial institutions and the like. Although they could hire local talents to handle some of these issues, many of these activities have to be performed interactively between foreign subsidiaries and headquarters (Luo & Tung, 2007, p. 495).

Cooke (2008) identified overall problems confronting Chinese MNCs regarding management and strategic decisions, such as the shortage of skills among the Chinese managers in handling Merger and Acquisition (M&A) negotiation and post-M&A integration, lack of knowledge of the business environment of the host countries and international standards. In the HRM area, one of the most difficult aspects is the necessity to manage a diverse workforce as a legal obligation to enhance organisational performance. Most Chinese managers tend to have a weak level of awareness of labour standards. Chinese managers are also confronted with different norms, policies and procedures regarding HRM functions such as recruitment, performance management and reward systems, training and development, employee participation and involvement, welfare and well-being, and external engagement with trade unions, government labour agents and judicial systems. Generally speaking, a corporate HR strategy needs to be informed by the legal environment, labour market characteristics and cultural norms of the host country. The challenge for managing HRM in a global context is a generic one but each may face a unique institutional and cultural environment that shapes the management practices of each specific host country (Cooke,

2008, p. 260). Regarding the detailed challenges, Zhu et al. (2008) raised a number of key research questions, such as how does outward FDI affect HRM in Chinese firms and their integration of strategic HRM, and what are the differences between HRM policies utilised for domestic and overseas operations?

Cooke's (2012) study of two leading Chinese IT companies, namely Huawei and ZTE, identified a research gap with regard to how Chinese MNCs' business strategy may impinge on their HR strategy on the one hand, and how and to what extent the HR strategy may support and/or constrain the fulfilment of international business (IB) strategy on the other. In her research, Cooke identified that both companies were primarily market-seeking and resource-seeking (R&D capacity and close to customers) and thus needed to overcome particular challenges associated with this: both developed and deployed political and social capital to meet these challenges (Cooke, 2012, p. 1847). Their IB strategy reflected the Chinese traditional value of harmony in order to gain local and global acceptance and to avoid direct competition and confrontation. The companies' HR strategy played an important role in supporting their IB strategy. However, both companies did not have an integrated global HR strategy that was supported by a sophisticated HR system from their headquarters. Instead, it was a mixture of HR practices transferred from the parent company, initiatives developed by managers at the subsidiary level and perceived best practices adapted from Western MNCs. In addition, both companies made an effort to train HCNs in China and Chinese employees aboard in order to develop social capital to support their business and to promote shared goals and cultural values between HCNs and PCNs (Cooke, 2012, p. 1847).

Along these lines, Warner and Nankervis (2012) analysed the HRM systems of three Chinese MNCs, namely Haier, Huawei and Lenovo based on published public information, including scholarly articles and internet sources. They argued that the rhetoric of modern HRM architecture is evident in these companies with organisational cultures reflecting their home culture context. However, the reality is that, in the process of their global expansion, they might have to adapt to the new international environment with a higher degree of localisation. With respect to the concept of "strategic choice", these companies displayed different emphases in their approaches to employee influence, human resource flow, reward systems and work systems, and these were associated with the influence of the operating country's systems. For example, Haier focused on innovation, equity and individualisation, with accordant employee recognition and motivational programmes. In contrast, Huawei emphasised teamwork, loyalty and employee commitment, with associated communication channels and competitive remuneration, and adopted decentralised HRM functions. Lenovo developed more advanced HRM strategy including talent management, human resource development, leadership, performance management and retirement benefits programmes, together with corporate social responsibility and employee health and safety projects. Therefore, their HRM practices might be at odds with their universalist rhetoric (Warner & Nankervis, 2012, p. 78).

Cooke and Lin (2012) furthered the empirical research into the investment motivation, strategy and labour management practices of Chinese MNCs operating in developing countries, such as Vietnam. By using three case study firms in the construction and light manufacturing industries, they investigated the institutional supports for and barriers to these state-owned and privately owned firms from both home and host countries. The key reasons for the two low-tech and labour intensive companies to invest in Vietnam were labour shortages and the rising cost of wages in Guangdong, China. The perceived low wage cost and a consumer market for cheap products in Vietnam were the attraction for the owners to make the investment decision. As for the construction state-owned enterprise (SOE), it had many years of overseas business expansion, and over the last two decades the company had bid for many infrastructure projects in Vietnam, building bridges, roads, power stations and public facilities. A number of challenges these companies were confronting included strict labour regulations, such as the Labour Code and the Trade Union Law; rising wage cost and strike actions; different work-life values; political resistance; skill shortage; government support (or lack of support) in both countries; and lack of company resources, particularly financial capital, technological competence and managerial capability. Cooke and Lin (2012) summarised two key findings, first, Chinese MNCs need political, cultural and social capital as well as other organisational resources to develop their operations overseas; and second, actors in the global value chain and national governments play an important role in shaping Chinese MNCs' operational environment and management practices.

Fan et al. (2012) adopted cross-level and in-depth interviews to extend empirical research to identify factors affecting global integration of Chinese MNCs in Australia. They found that the majority of factors (22 out of 26) the IB literature argues affect the global integration of MNCs were relevant to Chinese MNCs. However, the study revealed that several factors were not relevant to Chinese MNCs in Australia, for example, advertising intensity and premium positioning were not deemed important by senior executives. It also disclosed that senior management in Chinese MNCs emphasised building effective managerial channels such as information flow, control flexibility and resource flow between HQ and subsidiaries, as well as achieving managerial efficiency such as cost reduction, standardisation and economy of scale. Most Chinese MNCs did not claim they had country-specific advantages (CSAs), but had some firm-specific advantage (FSAs) such as leading technology in some fields. This study claimed that Chinese MNCs had the managerial abilities needed to transfer know-how to advanced economies and to obtain competitive advantage through global integration (Fan et al., 2012, p. 22).

Thite et al. (2012) developed a conceptual framework of global HR strategies and practices in MNCs from emerging economies across their subsidiaries in both developed and developing markets. They found there were different patterns among these MNCs operating in developed and developing economies. For instance, when they invested in developed economies, these emerging MNCs adopted M&A as the preferred method and the majority of them would have fully controlled ownership. Their motivation was to obtain technology, raw materials, skills and expertise and the local infrastructure to support their trade. In terms of their IHRM strategy, they preferred an adaptive and polycentric approach with a localisation orientation. In contrast, when they invested in developing economies, these emerging MNCs preferred greenfield investment through joint ventures. Their IHRM strategy was more associated with the ethnocentric approach that involved a wholesale transfer of the parent firm's systems, policies and personnel. Their case studies of Indian IT companies demonstrated that organisational structure and systems emphasised organisational culture and were codified in a manual throughout their global operations. They also paid attention to training leaders globally and decentralisation of management at the subsidiary level. Their IB strategy was based on an entrepreneurial spirit to exploit the opportunities ahead of their competitors. In addition, their global staffing strategy was to have the majority of positions filled by locals with senior management positions filled by PCNs. Their HR systems, such as recruitment and selection, career management and performance management, were similar across the global operations with the flexibility to accommodate local laws and HR trends and practices (Thite et al., 2012, p. 256).

Zheng (2013) challenged the existing strategic IHRM frameworks and discussed their inadequacy when applied directly to emerging MNCs from the so-called BRICS countries, including China. Some key implications were identified in this study: first, existing literature on emerging

MNCs tended to focus on global strategy and international business and downplay the link to evaluate the effectiveness of the cross-border people management practices of those firms. It should be useful to document and develop new knowledge to facilitate further management training for people from emerging MNCs and to harness good practices among them. Second, emphasising both external and internal drivers for internationalisation and emerging MNCs must actively influence new policy measures. Third, the key challenge for emerging MNC managers would be to design IHRM functions and activities that fit the laws, culture, society and politics of both home and host countries and are flexible enough to attract, develop and retain talent. In addition, emerging MNC managers must emphasise the outcomes associated with not only company's competitiveness but also wider societal and global village concerns (Zheng, 2013, p. 12).

Wang et al. (2013) investigated the influence of personality traits on cross-cultural competence (CCC) in Chinese expatriate managers' overseas managerial experiences. Two key research questions were asked: what personality traits contribute to CCC of Chinese expatriate managers, and how do these traits contribute to CCC under the influence of home and host institutions? They targeted 25 Chinese expatriates, 15 their foreign colleagues and 10 other people including expatriates' supervisors or cross-cultural consultants. Their interesting findings showed that two major personality traits contributed to the CCC of Chinese expatriate managers, namely conscientiousness and openness. Both positive and negative elements are associated with these traits. For example, being responsible is an important element of conscientiousness, but Chinese expatriates operating in different countries had a different focus.

Generally speaking, in developed economies, Chinese expatriates claimed that being responsible helped them comply with local regulation and fulfil expectations of local business partners. It also helped gain trust from local business partners/employees. However, in developing economies, being responsible mainly focuses on the deal with tough physical environments. In addition, there were negative elements associated with being responsible, such as Chinese expatriates paid more attention to relationship building and "face" culture and were not strict with the contract, plans and other rules. These could create conflict between Chinese expatriates and their foreign counterparts. Another issue is being diligent or working hard among the Chinese managers. Many Chinese managers worked very hard at their foreign subsidiaries and they expected the local people to do the same, such as taking overtime work. Most foreign managers and employees did not want to do so except the Japanese counterparts in Japan. This reflects the different social contexts which influence the behaviours of their foreign colleagues. The second personality trait was openness. Many Chinese expatriate managers commented that one important lesson was that a successful expatriate must be willing to learn new things. However, many Chinese expatriate managers were not willing to listen to different ideas due to their intention to preserve "face" in front of other colleagues (Wang et al., 2013, p. 3824).

Fan et al. (2013) recently conducted empirical evidence-based research into the IHRM strategies of three Chinese MNCs operating in Australia, namely one banking company, one plastic manufacturing company and one resources production company. There are a number of interesting findings from this qualitative research project. First, these cases represented different Strategic International HRM models: the banking company had a transnational strategy to promote both global integration and local responsiveness, the manufacturing company had a multi-domestic strategy to emphasise on local responsiveness and the resources company had an international strategy characterised as "low global integration and low local responsiveness".

Second, each company adopted different Strategic International HRM models which had an associated number of different characteristics. For instance, the banking company experienced a significant transformation with a constant interplay between organisational dynamics, such as structural context, institutional conditionality and control flexibility (Fan et al., 2013, p. 532). One major challenge under the old system which needed to be transformed in order to attract the best local employees was its outdated performance management and reward systems. Most banking management positions were filled with Chinese expatriates and their salaries had been gradually improved to match their local counterparts. However, there were still a number of weaknesses in the area of IHRM such as lack of mechanism to fully develop employees' potential and lack of international experience among Chinese managerial staff.

On the other hand, the manufacturing company successfully merged with one of the leading Australian plastic production companies and its strategies emphasised localisation policy with empowerment of local managers (with Australian managers managing business as usual), decentralisation of managerial roles and offering local managers flexibility to the greatest extent. Therefore, its international business strategy was mainly to facilitate local responsiveness and localise its operations in the host country. Accordingly, its IHRM policy focused on promoting autonomy, building a trust relationship with local executives, valuing local human resources and enhancing knowledge transfer back from overseas subsidiaries to domestic subsidiaries and facilitating cultural integration (Fan et al., 2013, p. 534). The authors claimed that many Chinese MNCs adopted this model and the core idea was to learn by participating while relying on local managers operating the business. The headquarters adopted the so-called *wu wei* (active no-action) through the "invisible leadership" to overcome their HR constraints.

Being different from the last two examples, the third company in the resources industry adopted a centralised IHRM model. The reason for investing in Australia was partly due to the similarity of mine ores in both host and home countries. Its international strategy was to exploit the parent company's knowledge and capabilities through worldwide diffusion. However, such centralised policy faced a number of problems, including cultural distance and conflicts between Chinese expatriate managers and local employees, given the senior level positions were usually appointed by the headquarters in China. The later challenge was how to make an effort to transfer effective HRM practices used at home to their host-country subsidiaries. One possible way was to adopt the "best practices" that have been implemented in both countries (Fan et al., 2013, p. 536).

Generally speaking, these examples demonstrate that Chinese MNCs are confronted with the constraints of international human resource deficiency. The overall situation is that China lacks personnel who possess international management skills, sufficient knowledge about market conditions of host countries and a good understanding of the intricacies of international business (Fornes & Butt-Philip, 2011). In the following sections, we will present our case studies regarding their cross-cultural management and HRM practices and then discuss the relevant implications.

Methodology

The context

Generally speaking, the institutional environment, cultural norm and way of life in Australia is very different from China and that could be seen as the major challenges for Chinese managers to successfully manage cross-cultural issues in general and HRM in particular. Chinese managers face many difficulties when they try to adopt new approaches and encounter certain obstacles from time to time. Therefore, these Chinese managers must be able to overcome difficulties with self-determination, proactive thinking and action, self-control and persistence, and be able to implement their strategy effectively (Li et al., 2012). We therefore selected Chinese managers working for Chinese MNCs, who had been working in Australia for a number of years with an objective to improve themselves and make a difference in the daily management activities.

Methods adopted

Using a qualitative ethnographic (case study) approach (Morgan & Smircich, 1980) for data collection, five experienced Chinese managers from industries of electrical appliance, trade, airlines, property and mining were purposefully selected. In-depth interviews of these managers were conducted to develop case studies of each enterprise. The case studies developed can thus be considered as socially constructed perspectives of interviewed Chinese managers operating in Australia.

A qualitative case study approach was used because it allows for intensive research of new topic areas in a manner that is not restricted by limited or narrowly defined variables (Eisenhardt, 1989; Eisenhardt & Graebner, 2007). This methodology is also considered appropriate for addressing research questions rather than testing hypotheses (Edmondson & McManus, 2007; Yin, 2008). Therefore, the socially constructed case study is both theoretically and practically appropriate as the research questions being investigated are in a relatively new area of Chinese managers managing cross-cultural issues and HRM in a totally different social, economic, institutional and cultural environment (i.e. from China to Australia). While in-depth qualitative analysis to develop significant insights happens at a later stage, researchers are able to get a sense of the data at the collection stage (Miles & Huberman, 1994). Researchers felt they had reached saturation point, in terms of new ideas, whilst collecting data on the fourth and fifth Chinese managers. Therefore, five cases were deemed sufficient to generate necessary understanding of the relevant issues.

The interviews were retrospective, semi-structured and in-depth in nature. The approach was similar to a collection of oral histories. The interviews conducted were in Chinese and were translated into English, accurately representing the meaning. The analysis of the data involved coding the responses of the interviewees. The NVivo software programme was used to code the data and generate themes that were extracted and clustered across the data set among these cases (Miles & Huberman, 1994, p. 87). While some codes that were identified relate to the existing literature on cross-cultural management, some codes were allowed to emerge organically during analysis, in particular related to different ways of handling things. For example, several codes such as relevant experience, relevant education, learning on the job, business plan, perseverance and social network that were identified in the case studies can be found in the mainstream cross-cultural management literature. However, there were some codes that were identified in these case studies that could be considered as unique to these Chinese managers such as "personal emotion" and "face". This initial round of analysis produced several codes that were then put under themes that were also motivated by the extant literature. These codes were initially broadly categorised into themes that could be related to (1) the overall quality of HR management in a cross-cultural environment; (2) enterprise cultures (that is, adopting cultural elements of both home and host countries and combining them adequately); (3) the companies' capability to be familiar with and fit into the local industry norms and regulatory environment; and (4) handling cultural and behavioural differences between Chinese managers and local employees. The outcome of later rounds of analysis is presented as empirical findings in the next section.

Our case studies

The quality of leadership in general and HR management in particular have always been seen as challengeable issues. The leadership arrangement of HR management can be different among companies, in particular related to the size of a company. Larger companies tend to have an HR department with a professional HR manager leading a team managing HR issues, but smaller companies tend to allocate the HR tasks to a General Manager or Deputy General Manager who also has other administrative duties. Among the majority of Chinese companies, HR has been seen as one of the strategic areas and normally companies would put their own Chinese managers in charge of HR management.

Two important areas of Chinese MNCs' HR management need to be considered: (1) managing the Chinese expatriates in strategic positions in their subsidiary companies and (2) managing local Australian employees and engaging other local stakeholders such as local government agents and trade unions.

Regarding expatriates management, most Chinese expatriate managerial staff members were selected to take the positions of general manager, chief accountant, marketing manager and other business operational managers (or frontline operation managers). The selection criteria and performance indicators are different among managers with different positions. All those managers must be able to communicate in English, pass the selection examination and have previous overseas work experience.

However, the major problem was that many of these managers were not familiar with Australian law and regulations as well as other necessary knowledge of operating business in Australia. There was inadequate pre-departure training for them and these managers were basically learning on-the-job during the expatriation period.

Those Chinese managers claimed that there has been an improvement of expatriates' salary in recent years, in particular most Chinese MNCs have adopted host-country salary standards rather than home-country-based salary plus allowances. Many managers can also take their spouses with them to Australia which was not the case in the past. However, all the expatriate managers are required to take on multiple duties and there has been an increase in the number of key performance indicators (KPIs) – some managers have more than 20 KPIs and most of them are linked with cost control and profit margin.

As for managing local Australian employees, HR managers are confronting a number of problems. Most local employees were recruited through a local HR agent. However, most Chinese MNCs found that using an HR agent for recruiting employees could cost more than A$3,000 for one person and they did not have the budget to cover this. Therefore, in recent years, they would rely on the introduction of colleagues and friends as well as using online advertising tools to recruit local employees. On the other hand, due to the salary budget being under the central control of headquarters as part of cost control measurement, the salary payment for the local employees was not high enough to attract capable employees with good experience. The consequence was that new employees performed very well during the interview but they could not meet expectations when they were working on a particular task. One General Manager (airlines) complained:

> It is very hard to dismiss incapable employees in Australia. They have a lot of protection from Union and other regulations. We have to accumulate enough evidence and pass many hurdles in order to dismiss someone who is not performing well.

Other employees with more experience were normally relatively older who were less productive but able to act as mentors and train younger employees. In recent years, given there has been an increasing number of young Chinese skilled migrants and university graduates settling down in Australia, many Chinese MNCs have started to recruit new staff and arranged on-the-job training provided by those older employees with experience. One HR Manager (trade) commented:

> These young Chinese employees work very hard, but they need to learn more with our older local employees in order to master the operation process. They have good attitude towards work and hardly take any leave. The communication is much easier because they speak both Chinese and English. They also feel fortunate to have a job and show their gratitude.

This comment shows the contrasting attitude towards local Australian employees expressed by the Chinese managers. They claimed that one of the major challenges of managing local Australian employees was dealing with "attitude" towards work. For instance, a General Manager (airlines) complained about his Australian employees' taking frequent leave with dubious reasons of false family problems or sickness:

> Australian employees always take leave with make-up reasons such as handling family problem or feeling ill. This has negatively influenced on our daily work significantly. They believe that taking leave is their legitimate entitlement and they must use them fully. This is very different from our Chinese employees in China.

Another example is related to the different understanding of commitment towards work and organisation. An HR manager (electrical appliance) complained about his Australian employees:

> Australian employees don't take extra care about their work. When the time reaches 5pm, you can't expect them to stay longer working for the company. They all rush to home at that time. This is very different from our Chinese employees.

One of the constraints of Chinese MNCs to motivate and reward local employees is that the total budget is under the control of their headquarters in China and they cannot spend more on the salary budget to pay bonuses or other rewards – even their overall performance is better than the initial key performance indicators (KPIs). Professional training is also limited due to minimising cost measurements. Many local employees do not think these are good policies and they tend to change their job when the opportunity arrives. Therefore, lack of incentive pay and professional training can also lead to the problem of retaining capable employees among Chinese MNCs, and the loyalty and commitment among the Australian employees towards the Chinese MNCs are generally low.

Most Chinese MNCs try to do their best to build productive and effective relationships with external stakeholders, in particular the government agents and trade unions. A number of our sample companies had union branches. The others did not have union branches, but a number of employees belonged to their occupation union. All these Chinese managers claimed that it is important to have a cooperative relationship with unions and they normally worked with unions for dispute settlement and negotiation on other labour issues. However, most of them admitted that they knew little of local labour law and regulations, and in many cases, they had to rely on local labour lawyers to handle the dispute cases with huge cost.

We observed there were a number of consistent issues between previous research and our case studies. For example, among these Chinese MNCs (e.g., manufacturing) who adopted M&A as preferred methods, their IHRM strategy would be an adaptive and polycentric approach with more localisation orientation. Those services-oriented businesses (e.g. trade, airlines and property) with wholly owned capacity normally adopted centrally controlled HRM practices with some key characteristics such as most senior managers were selected from China; the average wage level was relatively lower in comparison with the local counterparts; there was no orientation training; they were not familiar with local IR and HRM systems; and they lack of cross-cultural management experience. As for other larger mining Chinese MNCs, they normally adopted mixed HRM policies and practices by combining the so-called best practices in Australia and China. In terms of the senior management team, they normally combine both local and Chinese expatriates, in particular because, according to the Australian law, the company CEO and CFO must be local Australian citizens. Even in this kind of company, cross-cultural communication was a challenging issue given a number of Chinese managers would keep their own management mentality and their own way of managing things in Australia that would not be appropriate in a different cultural site. Areas such as environment protection, respecting law and due diligence on organisational health and safety (OH&S) could be compromised among the Chinese managers due to lack of awareness and experience dealing with these sensitive issues abroad.

Summary

From the previous research findings, as well as our own case studies, we can see that a number of factors influence Chinese MNCs' decisions about whether to adopt appropriate IHRM policy and practices. Given the overall picture, the quality and capability of Chinese managers are not well equipped in terms of cross-cultural communication and management, unfamiliar with local laws and regulations, lack autonomy to make key decisions in the financial and HR areas and have an ethnocentric way of carrying on in their own management styles and attitudes (e.g. face culture and relationships are more important than "rule of law"). The fundamental question people might ask is: "Are Chinese companies ready for offshore business investment?" Our answer is "not yet". There is so much homework to be done and so many issues still challenging their routine operation. We think that providing adequate pre-departure training for Chinese managers working overseas is one of the fundamental things Chinese MNCs should do (Zhang & Fan, 2014). A period of overseas practice for these managers before taking management position in an overseas operation is also crucial for accumulating experiences. Chinese MNCs should also enlarge their talent pool by including overseas trained and educated Chinese professionals into the potential management team and be vigilant about identifying talent targets in the overseas market. This also requires a large number of state-owned enterprises giving up the traditional political appointment of their senior managers in their organisations. Overall, we find more localised HRM policy and practices which not only are better than centralised ones but also provide local HR participation as well as fully utilising local talents in many areas, including know-how, local business network, familiarity with local law and regulation and, most importantly, adopting appropriate local cultures and norms. We feel this is another "long march" for both Chinese government and companies who are engaged in overseas business operation to realise their potential and eventually make Chinese overseas business operation successful.

References

Cooke, F. L. (2008). *Competition, Strategy and Management in China*. Basingstoke: Palgrave Macmillan.

Cooke, F. L. (2012). The globalization of Chinese telecom corporations: Strategy, challenges and HR implications for the MNCs and host countries. *The International Journal of Human Resource Management, 23*, 1832–1852.

Cooke, F. L. & Lin, Z. (2012). Chinese firms in Vietnam: Investment motives, institutional environment and human resource challenges. *Asia Pacific Journal of Human Resources, 50*, 205–226.

Edmondson, A. C. & McManus, S. E. (2007). Methodological fit in management field research. *Academy of Management Review, 32*(4), 1155–1179.

Eisenhardt, K. M. (1989). Building theories from case study research. *Academy of Management Review, 14*(4), 532–550.

Eisenhardt, K. M. & Graebner, M. E. (2007). Theory building from cases: Opportunities and challenges. *Academy of Management Journal, 50*(1), 25–32.

Fan, D., Zhang, M. M., and Zhu, C. J. (2013). International human resource management strategies of Chinese multinationals operating abroad. *Asia Pacific Business Review, 19*, 526–541.

Fan, D., Zhu, C. J., & Nyland, C. (2012). Factors affecting global integration of Chinese multinationals in Australia: A qualitative analysis. *International Business Review, 21*, 13–16.

Fornes, G. & Butt-Philip, A. (2011). Chinese MNCs and Latin America: A review. *International Journal of Emerging Markets, 6*, 377–397.

Li, J., Young, M. N., & Tang, G. (2012). The development of entrepreneurship in Chinese communities: An organizational symbiosis perspective. *Asia Pacific Journal of Management, 29*, 367–385.

Luo, Y. & Tung, R. (2007). International expansion of emerging market enterprises: A springboard perspective. *Journal of International Business Studies, 38* 481–498.

Miles, M. B. & Huberman, A. M. (1994). *An expanded sourcebook: Qualitative data analysis* (2nd ed.). London and New Delhi: Sage Publications.

Morgan, G. & Smircich, L. (1980). The case for qualitative research. *Academy of Management Review, 5*(4), 491–500.

Scullion, H. (1994). Staffing policies and strategic control in British multinationals. *International Studies of Management and Organization, 24*, 86–104.

Shen, J. (2004). Compensation in Chinese multinationals. *Compensation and Benefits Review, 36*, 15–25.

Shen, J. (2005). Towards a generic international human resource management (IHRM) model. *Journal of Organizational Transformation and Social Change, 2*, 83–102.

Shen, J. (2006). Factors affecting international staffing in Chinese multinationals (MNEs). *International Journal of Human Resource Management, 17*, 295–315.

Shen, J. & Edwards, V. (2004). Recruitment and selection in Chinese MNEs. *International Journal of Human Resource Management, 15*, 814–835.

Thite, M., Wilkinson, A., & Shah, D. (2012). Internationalization and HRM strategies across subsidiaries in multinational corporations from emerging economies – A conceptual framework. *Journal of World Business, 47*, 251–258.

Wang, D., Freeman, S., & Zhu, C. J. (2013). Personality traits and cross-cultural competence of Chinese expatriate managers: A socio-analytic and institutional perspective. *International Journal of Human Resource Management, 24*, 3812–3830.

Warner, M. & Nankervis, A. (2012). HRM practices in Chinese MNCs: Rhetoric and reality. *Journal of General Management, 37*, 61–80.

Yin, R. (2008). *Case study research: Design and methods* (4th ed.). Beverly Hills, CA: Sage Publishing.

Zhang, M. M. & Fan, D. (2014). Expatriate skills training strategies of Chinese multinationals operating in Australia. *Asia Pacific Journal of Human Resources, 52,* 60–76.

Zheng, C. (2013). Critiques and extension of strategic international human resource management framework for dragon multinationals. *Asia Pacific Business Review, 19,* 1–15.

Zhu, C. J., Thomson, S. B., and De Cieri, H. (2008). A retrospective and prospective analysis of HRM research in Chinese firms: Implications and directions for future study. *Human Resource Management, 47,* 133–156.

7

Corporate Sustainable Development: How and Why Chinese-Invested Firms Engage Community in the Australian Mining Industry

Introduction

Corporations are part of our society and can play a significant role in sustainable development because of their "economic scale and significance" (Clarke, 2007, p. 223) and use of productive resources. Despite the potential role played by business communities, sustainability beyond the enterprise or institutional level imposes major challenges to corporations for several reasons.

First, as the demand from the general public, customers and communities for corporations to take responsibility in both social and ecological areas rapidly grows worldwide, governments and regulators have increasingly viewed sustainability issues more seriously and are gradually promulgating laws and regulations that require all corporations to perform certain activities in the area of sustainable development.

Second, institutional investors have taken into consideration the sustainability performance of companies, both legally and morally. Thus, they are more inclined to finance businesses with sound sustainability performance.

Third, organisations that take corporate sustainability seriously are more attractive to employees (Schmidt & Freeman, 2000; Turban & Greening, 1997) and, being an employer of choice, have considerable impacts upon its reputation.

Finally, local communities and the general public can have their concerns and voices heard quickly and widely through the Internet and increasingly powerful social media such as Twitter, Facebook and WeChat. Hence, it is easy for these communities to ally with media and relevant stakeholders in other parts of the world to impose pressure on corporations. The pressure, either legally, economically or morally, both externally and internally, has been rapidly growing on corporate executives and directors to consider sustainability issues as crucial to the survival of their corporation. In other words, corporate sustainability reporting has been regarded as "the de-facto law for business", according to a survey conducted by KPMG (2011).

The move for corporations to pursue corporate sustainability not only requires substantial changes in their organisational values and purpose, but also demands them to develop new strategies and implement them successfully to incorporate social and environmental considerations into their business decision-making and operations. This has imposed challenges on many organisations as they need to understand how to implement these initiatives and develop organisational capabilities to undertake major strategic changes required in their organisation. Failing to do so would result in corporate sustainability becoming "greenwash" or "window dressing" (Slack, 2011).

Community engagement is a vital part of corporate social responsibility (CSR) and has also become an essential part of operations in the global mining industry. CSR has now become a "social license to operate" (SLO) (Owen & Kemp, 2013) in the mining industry because of the nature of mines operating near or within distinct geographic communities. More specifically, community engagement is the social dimension of CSR and often considered the most challenging aspect of CSR as it is deeply embedded in the cultural and institutional environment in the host country (Marquis & Battilana, 2009). Moreover, mining has substantial impact on the environment and society due to the nature of its operations, such as mining, minerals processing, use of land and water and traffic hazards caused by frequent long-distance transportation from mine sites to ports, and thus has a reputation of being a "dirty business" (Gifford et al., 2010, p. 304).

The mining industry has received increasing scrutiny from international institutions (e.g. the United Nations and OECD) and NGOs. Moreover, a number of industry-specific initiatives such as the Extractive Industries Transparency Initiative (EITI) and reporting practices (e.g. Global Reporting Initiative [GRI]'s sector supplement: Minerals and Metals) have been launched to regulate corporate sustainable practice in the mining industry. In addition, industry associations, such as The International Council on Mining and Metals (ICMM) and

Minerals Council of Australia (MCA), have developed industry-specific sustainability initiatives, such as the Sustainable Development Framework, and Enduing Value, to guide its members in practicing their corporate sustainability. Mining companies have limited mobility and must operate where the resources are located (Ellem, 2003) which may prompt them to engage more seriously with sustainability initiatives and with the communities in which they operate (Jones et al., 2007).

This chapter examines the community engagement by a Chinese-invested mining company in Australia. It aims to address the following research questions: (1) What community engagement strategy has been undertaken by the mining company? (2) How does the company conduct its community engagement activities and programmes? (3) Why does the company engage communities in their business operations?

This research is important for several reasons. First, community engagement in the Australian mining industry has a long history, as mining has potential impact on local communities in many ways – economically, environmentally and socially – and has advanced globally over the past two decades (Minerals Council of Australia, 2005). Second, Australia has a long mining history and has relied heavily on its mining industry for economic prosperity and employment (Drysdale & Findley, 2008). Third, community engagement has become an important part of establishing and maintaining legitimacy, frequently called a social license to operate (SLO) (Owen & Kemp, 2013). Finally, Chinese firms have invested heavily in the Australian mining industry (Ministry of Commerce, 2014). It is not clear how they perform from a CSR perspective and what challenges they have in practicing their CSR abroad.

A conceptual framework of CSR performance

Much research has been devoted to CSR by multiple disciplines over the past half-century since the publication of Bowen's (1953) work in this field. The dominant theme in the literature so far has been CSR motivation and performance, and relationship between CSR performance and organisational financial or non-financial performance, which helps understand *why* firms engage in CSR activities and *if* they benefit from such activities.

More specifically, an overwhelming proportion of these studies have focused on corporate social performance (CSP) and its link or "coupling" (Lee, 2008) with organisational performance (Barnett & Salomon, 2006; Orlitzky et al., 2003). Moreover, these studies have identified a number of conditions under which CSR performance affects organisational performance, such as profitability (Brammer & Millington, 2004), slack resources (Waddock & Samuel, 1997), CSR task visibility (Jiang & Bansal, 2003), corporate governance and R&D investment. Reasons for the influence of CSR performance have been determined by studying the key mechanisms and processes, such as stakeholder influence capability (Barnett, 2007) and managerial interpretation of CRS (Sharma, 2000). These studies have shed considerable light on corporate sustainability practices. However, literature in CSR has been very fragmented, as shown by a recent extensive review of 588 journal articles and 102 books published since 1970 (Aguinis & Glavas, 2012).

In this section, we propose a conceptual framework of CSR performance based on the review of CSR literature, as well as by focusing on CSR motivations, firms' CSR strategy and resources and CSR context.

A number of CSR motivational factors have been identified in the literature. Institution, stakeholder pressure and ethical motives (mainly driven by top management team members and company values) have been widely considered as major driving forces (Banerjee et al., 2003; Bansal, 2005).

Stakeholder theory. Stakeholder theory (Freeman, 1984) has been widely used to identify what type of stakeholders drive or inhibit firms to adopt sustainability practices. Consequently, a number of key stakeholders have been identified, including government, communities, the public, media, top management, shareholders, customers, suppliers, or lobby groups (Henriques & Sadorsky, 1996; Sharma et al., 1999).

The essence of stakeholder theory is that relationships with key stakeholders should be managed to create value for the company (Parmar et al., 2010). Stakeholders can supply key resources to the company, and have sufficient power to influence, either positively or negatively, the company's performance (Post et al., 2002). Thus, they can contribute, either positively or negatively, to the company's wealth-creating capability and activities. The relationship with key stakeholders can determine a company's capability to generate sustainable wealth and long-term success (Post et al., 2002). More specifically, stakeholder management can enhance an organisation's reputation and financial performance, attract potential partners and employees and reduce its risks (Parmar et al., 2010). Communities are an important type of stakeholder for companies, although not all communities are stakeholders, particularly primary stakeholders. Stakeholder theory at the organisational level provides insight into why and how companies engage communities.

Donaldson and Preston (1995) suggested there are three approaches to managing stakeholders: instrumental, normative and descriptive. The instrumental perspective to stakeholder management emphasises the link between stakeholder management practice and the achievement of a company's performance. Its key question is how stakeholder management can improve or deteriorate that performance. The normative perspective explains how stakeholder relationships should be managed based on the moral or philosophical orientation of the company. This is in line with the ethics perspective, a point which will be elaborated later. The descriptive perspective aims to enhance our understanding about how stakeholder management is practiced in organisations, rather than elaborating why stakeholders can or should be managed in the first two perspectives.

Stakeholders can be managed based on some of their characteristics, such as power, legitimacy and urgency (Mitchell et al., 1997). Communities, particularly those located near the company's operations, are often considered as primary stakeholders, along with customers, suppliers, channel distributors, shareholders, financial providers and employees. From a social network perspective, understanding the position of the firm and community in the network is also important (Rowley, 1997) as this can influence the strategies used to manage firm-stakeholder relationships. The emphasis on stakeholder analysis is at the organisational level.

Institutional theory. Institutional theory (Peng, 2003; Wright et al., 2005) is another theory that is widely used in understanding why businesses undertake CSR. Institutions are the "rules of the game in the society" and consist of formal rules (constitutions, laws, property rights), informal constraints (sanctions, taboos, customs, traditions, codes of conduct) and characteristics of enforcing these constraints (North, 1991; 1992). Therefore, institutions "structure incentives in exchange, whether political, social, or economic" (North, 1992, p. 477). Institutional theory has been widely used in the studies of why and how companies respond to their institutional environment to seek legitimacy (Bondy et al., 2012; Scott, 1987; Suchman, 1995). Gaining legitimacy is crucial to companies as legitimacy is the acceptance and approval of organisational actions by external constituents (Kostova et al., 2008, p. 1001) and its acquisition may provide access to critical resources or act as a safety net for mitigating risks.

While the notion of institution has three dimensions: regulative, normative and cognitive (Scott, 1995), the legal responsibilities primarily come from the regulative dimensions – more specifically, regulations and laws from the governments at both national and international levels – or "macro-institutional pressure" (Bondy et al., 2012, p. 282). Firms have to conform to laws and regulations at a minimal level to be considered legitimate.

The cognitive element of a national institutional environment describes the widely shared cognitive categories used by people, including business executives and communities. These categories include the level of economic development, role of government and business and ideology such as socialism and capitalism. They will affect expectations of the role business should play in community development.

The normative aspect of national institutional environment is primarily about the culture in the host country, such as norms, beliefs and assumptions. National culture can influence the way firms engage with communities. For example, Japan is rated highly in terms of collectivism and Japanese companies tend to emphasise their internal employees, rather than communities (Wokutch, 1990); in contrast, Chinese companies have a narrow focus on CSR activities, often only on philanthropic donations. They lack CSR strategies and are weak in CSR implementation and communications (CNOOC, 2014).

Institutional theory is very useful in explaining why companies implement CSR. Although traditional institutional theory (Friedland & Alford, 1991) has emphasised the importance of context, such as law and regulations, culture, values and norms, the neo-institutional researchers (i.e. DiMaggio & Powell, 1983) have concentrated on the fundamental issue of why organisations in an organisational field structure behave similarly. The concern over acquiring legitimacy drives companies operating in a similar external institutional environment to adopt similar organisational structures and practices in order to conform to regulatory requirements and social acceptance and to gain creditability. This renders organisations in an organisational field isomorphic with each other. Three mechanisms have been identified which drive the outcome of similar organisational behaviours: coercive, mimetic and normative (DiMaggio & Powell, 1983). Bice (2013) draws upon key new institutional theory concepts to offer pertinent insights into the formulation of community engagement policies and practices at a local level to understand community activities in the Australian mining sector.

Ethics theory. At an individual level business ethics provide a philosophical foundation for examining the morality of decision-making.

Ethics offer prescriptive guidance for individual decision-makers when faced with ethical dilemmas on how they ought to act (Trevino & Nelson, 2014). Three ethical perspectives have been widely used in explaining how decisions can be made, namely, utilitarianism, deontology and virtue ethics. While utilitarianism theory considers utility as the foundation of morals and is focused on "the end" or consequences of actions, deontological ethics theory suggests ethical decision-making should be based on universaliable ethical principles (i.e. promise keeping or truthfulness) irrespective of the consequences of the decision (Bondy et al., 2012, p. 108). Simply put, it stresses the importance of the decision-making standards. Deontology and utilitarianism focus on the decision-making, while virtue ethics theory concentrates on the person making the decision. It emphasises the personal characteristics of the individual and encourages decision-making that contributes positively to the individual's acquisition of virtuous disposition.

Individual morals, particularly in senior management, are important in pursuing corporate sustainable development. This is what managers *want to do*. Top management can play a central role in corporate sustainable development through their direct involvement and demonstration of their commitment by appointing senior managers who are in favour of this development (Banerjee et al., 2003). Other individuals within the company, such as champions and employees, may also be important for corporate sustainable development (Bansal & Roth, 2000).

Context. Several key contextual variables have also been identified in the extant literature, including: country effect, industry type (field cohesiveness, issue salience) (Banerjee et al., 2003; Bansal & Roth, 2000), firm size (Darnall et al., 2010) and organisational context (Sharma et al., 2007). For example, the mining industry was labelled a "dirty" industry (Bansal & Roth, 2000) and has a close or "cohesive" network due to its negative image. Firms in a cohesive network are less likely to be motivated by pursuing higher levels of ecological responsibility (p. 731). Regarding firm size, smaller firms were more responsive to value-chains, internal and regulatory stakeholder pressures than their larger counterparts (Darnall et al., 2010).

Organisations' CSR strategy and resources. An organisation's CSR strategy and resources can also affect its CSR performance. Carroll (1979) has identified four CSR strategic stances firms can take: obstructionist, defensive, accommodative and proactive. These four stances have been widely used in CSR literature (Clarkson, 1995; Sauser Jr, 2005; Wartick &

Cochran, 1985). They are hierarchical and can be considered as a spectrum ranging from obstructionist or "reactive" (Wartick & Cochran, 1985) to proactive.

Following the approach used by Luo Shenkar and Nyaw (2002), we further classified these four stances into two broad strategies: defensive and offensive. The defensive approach covers reactive and defensive actions, and is developed to reduce operational and social risks, and secure legitimacy; in contrast, the offensive approach consists of accommodative and proactive actions and aims to creative value for the company (Porter & Kramer, 2011), improve reputation, maintain identity and enhance its competitive advantage.

Several studies have proposed that organisations can develop their organisational capabilities to respond to external pressures. For example, Hart (1995) has extended the resource-based theory to link a firm's capabilities to its environmental performance. He suggested that organisations develop their new capabilities that can facilitate environmental sustainable economic activity. These capabilities can be regarded as the sources of an organisation's strategy and competitive advantage in the future. Sharma et al. (2007) found that an organisation's capability in stakeholder engagement, continuous innovation and strategic proactivity affect its proactive environmental practices.

The drivers of corporate sustainable development and their key moderating factors can be depicted in Figure 7.1.

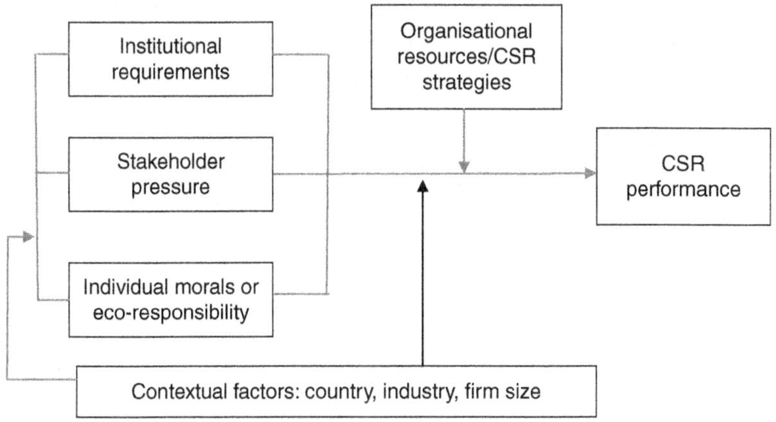

Figure 7.1 The key drivers and moderators of CSR performance

Research on community engagement

Much research has been devoted to corporate community engagement, particularly in the extractive industries (Eweje, 2006; Kemp et al., 2012; Muthuri et al., 2009). Two major themes of community engagement are the motivations for companies to engage with communities and community engagement strategy and implementation (Idemudia, 2009). The four broad types of motivation (institution, stakeholder, individual and corporate CSR strategy) outlined in Figure 7.1 are applied to corporate community engagement. In the following section, we discuss community engagement strategy and its implementation, which has formed the focus of a major stream of studies in community engagement.

Borrowing from leadership studies, Bowen et al. (2010) have suggested that community engagement strategies can be broadly grouped into three categories: transactional, transitional and transformational. These three types of strategies represent a continuum of community engagement ranging from "giving back", "building bridge" to "changing society".

Regarding the implementation of community engagement strategies in the mining industry, empirical studies have been quite fragmented and concentrated on several issues, including programmes or foci (Eweje, 2006), governance (Idemudia, 2009) and internal management (Kemp & Owen, 2013; Owen & Kemp, 2014). For example, Eweje (2006) found that a variety of community engagement initiatives has been undertaken by MNCs in Nigeria and South Africa, such as education, social welfare, infrastructure and small business development.

In Australia, community leaders describe these foci as the "flavors of the month" in terms the kind of engagement undertaken by mining companies (Huang & Staples, 2014). Sustainable community development has recently become a key theme in community engagement studies (Muthuri et al., 2012; Owen & Kemp, 2012). Idemudia (2009) examined the governance of community engagement programmes in the Nigerian oil industry by MNCs and found that two governance models have been used: an in-house community investment model and a corporate–community foundation model. He also found that the latter was better perceived by the local community as it not only involved communities and NGOs in decision-making, but also facilitated information-sharing and communication.

Several studies have recently investigated the internal mechanisms for implementing community engagement programmes (Kemp & Owen,

2013; Owen & Kemp, 2014). Kemp and Owen (2013; 2014) reported that community engagement has not been regarded as part of the core functions in many mining companies in terms of its roles, relationships with other departments in the company, its influence and inclusion in decision-making and professional status recognition. The researchers further suggested that four internal dimensions (external engagement, structural inclusion, access to resources and internal influence) can be used to understand the social performance of organisations in the mining industry.

Community engagement practices by Chinese firms

Most of existing studies on CSR have focused on companies from developed countries, particularly the United States and Europe. Although the concept of CSR has been developed since the 1980s in Western countries, it has only been recognised by Chinese firms and government in the last decade and is still in its developmental stage (Vermander, 2014). Nevertheless, the importance of CSR has gained recognition by the Chinese government, partly due to the rapid internationalisation of Chinese MNCs (Gugler & Shi, 2009). Consequently, the State-owned Assets Supervision and Administration Commission (SASAC) issued Guidelines to SOEs directly under the Central Government on fulfilling CSR in 2008 (SASAC, 2008). One reason for this is that

> [f]ulfilling CSR is the need for the Chinese SOEs to participate in international economic cooperation. As the progress of economic globalization, the international community concerns more and more on the performance of an enterprise in social responsibilities. By fulfilling CSR, it is either helpful in establishing a "responsible" public image by Chinese enterprises and more internationally influential, or significant for China to spread an image as a responsible nation.
>
> (SASAC's Guidelines, 2008)

As a developing country, China has also become a member of the UN Human Rights Council and has endorsed the UN Guidelines on Business and Human Rights (Zadek et al., 2012). China has also set up its targets for its sustainable development at the national level in its 12th five-year plan (2011–2015) and has issued many regulations in the last decade for environmental protection and social welfare, such as the Labor Contract Law issued in 2007 which specified a minimum wage.

Regarding corporate community engagement (CCE) in China, it is very different from that practiced by firms operating in Australia in many aspects, including the role of government and business, the developmental stage and focus. The role played by the Chinese government in CCE is very different. China is a socialist country and most of the large businesses are still state-owned. Thus, the government plays two roles: by providing social services and regulating business and by becoming the owner of business. Therefore, corporate conflicts are primarily handled by the government on behalf of business. Community engagement has not been perceived as critical to business operations in terms of risk management or legitimacy.

Regarding the focus and form of community engagement, the SASAC's Guidelines (2008) also state that Chinese SOEs should participate

> in social public welfare programs. SOEs ought to encourage their employees to volunteer for social services, and actively participate in community and social welfare program, such as charity, donations, and giving support to schools, cultural or health activities. Upon the occurrence of major natural disasters and emergency accidents, CSOEs also need to provide financial, material and manpower support.
>
> (SASAC's Guidelines, 2008)

Making donations to social welfare programmes, including charity, schools, cultural or health activities and natural disaster relief has been considered a major potential focus for CSR generally, or community engagement specifically. The more complex modes practised by Western business, such as infrastructure development, small business development and community development, have not been fully recognised, let alone practised, by Chinese firms in China. Community development is considered primarily the role of government as evidenced by the fact that the National Development and Reform Commission (NDRC) has set up a Division of Society Development responsible for community development, particularly initiatives in rural and remote areas.

Although CSR is on the rise and evolving quickly in China, and a number of large Chinese MNCs have started adopting the language and practice of sustainability (Zadek et al., 2012), it is still a relatively new concept to many Chinese companies. In addition, there are still huge gaps in terms of international management experiences, competences and HR resources, in order for Chinese MNCs to catch up with well-established MNCs in the global competitive arena.

Much research effort has been dedicated to the issues of corporate sustainability; however, many crucial research questions still remain unanswered. As Bansal and Roth (2000, p. 717) pointed out:

> Scholars have not detailed the conceptual distinctiveness of alternate ecologically based motivations, or established whether the categories of motivations are inclusive of all motivations or all mutually exclusive of each other. Existing research lacks clarity as to how motivations differ and what contexts lead to particular motivation.

Similarly, although community engagement research has been undertaken over the past two decades, a recent call for more "case-based studies" (Bowen et al., 2010, p. 312) was made after an extensive literature review on community engagement, due to the complex nature of community engagement such as divergent community definitions, various antecedents and different strategies used.

Background information on the company studied

The company under analysis is a joint venture (JV) between a large Chinese steel company (a state-owned enterprise [SOE]) and an Australian junior mining company engaged in exploration, construction and production of iron ore in the Midwest area of Western Australia. The JV was set up in 2006 on a 50–50 basis and has been primarily managed by an Australian team.

Many Chinese firms in the Australian mining industry rely on an Australian team to manage their business, particularly where the operations are large and complex (Huang & Austin, 2011). The project was originally budgeted at about A$1.8 billion in 2006 with a planned annual production capability of 16 million tons of concentrated iron ore more than 30 years. By the end of 2013, the project cost about A$2.7 billion, including the costs of building a 145-kilometre water pipeline, 180-kilometre high-voltage power line and 80-kilometre rail infrastructure connecting the mine site to the regional rail network and to the regional port for shipment. Geographically, these "linear infrastructures" have extended over five shires (one hosting, three transient and one terminal shire) and required the JV to negotiate with seven Aboriginal groups for access to their land to build the infrastructure and mining operations.

The mine is located 80 km away from the nearest town (the hosting shire) in the region and 200 km from the port. The main operational mode is fly-in-and-fly-out. However, a proportion of employees are drive-in-drive-out from local shires.

Research methods

A case study approach was adopted in this research so as to focus on the "why" and "how" questions relating to real-life events, with little interference from the researchers, making it an appropriate approach for such a research setting (Yin, 2003). Reliability and validity are often two major concerns which must be addressed in case studies (Gibbert & Ruigrok, 2010). Two approaches were used to ensure internal validity of this method. First, *multiple sources of evidence*, as suggested by Yin (2003), were adopted for this study to achieve triangulation. These sources of evidence include (1) interviews with three managers who are responsible for community engagement in the JV's headquarters followed by a number of phone calls and emails for further clarifications; (2) one field trip to the JV mine site to conduct interviews with seven senior local government officials and community leaders across two shires where the JV is operating and also to collect physical artefacts of community engagement; and (3) an analysis of the company website, the annual and sustainability reports. Second, the case study findings were sent back to one of the managers interviewed for validating.

The reliability of this case study has been assured through carefully designed data collection and analysis. A case study protocol was developed based on the research objectives and outcomes of literature review. Built on the research questions developed, the interviews were semi-structured and ranged from 50 minutes to two hours. All interviews were digitally recorded and then sent to an Australian professional service provider for transcribing. A list of codes was initially developed based on the outcomes of the literature review and the JV's annual and sustainability reports. Moreover, the suggestions by Yin (2003) to *create a database* for this JV was followed by placing all documents onto a shared drive where both authors could access and enter codes.

All the information collected was integrated using the qualitative data analytical techniques proposed by Miles and Huberman (1994), particularly *theme identification* and *time-order display* such as *event listing* and *activity record*.

External validity was sought by discussing the findings with managers working in other firms and community group leaders. As part of the requirements from the funding body, five research seminars were held in Australia in June–July 2014 with more than 400 participants. Feedback was sought from these participants during the seminar. No substantially different views on our findings were expressed during these three seminars.

Findings and discussion

Community engagement strategy and approach

Strategy. The strategic stance taken by the JV is primarily defensive for several reasons. First, from the Chinese partner's perspective, the JV is its only international operation. The Chinese partner has not been active in pursuing its international expansion strategy since establishing the JV in 2006. Clearly, it has little intention of using CSR as a competitive advantage in the global market. Rather, it is mainly focused on using community engagement to secure legitimacy. From the Australian partner's standpoint, the JV accounts for more than 90 per cent of its total assets. The Australian company is relatively small in size and has little resource to expand its business.

The community engagement strategy was developed after the JV conducted its external social impact assessment of the mining operations and internal analyses. The JV first approached local governments about community aspirations and held meetings with the general communities in the local town halls at the start of its mining project. It also recognised that it would take a lengthy time to move into cash generation (production) mode given the complex nature and large scope of its major project. Two principles were used to develop its community engagement strategy as the JV explicitly expressed it in its sustainability report:

> [C]ommunity expenditure needs to be carefully managed. To this end, the two driving principles of our community program are "targeted" and "cost-effective". We target our community investment into areas which are likely to deliver the most sustainable benefit.
>
> (Sustainability Report, 2012, p. 19)

Approaches. The JV has focused its community engagement activities on three strategic areas: "regional health, education, and youth development" (Sustainability Report, 2012, p. 19) based on the outcome of its initial assessment of community needs. Our interviews with the Community Relations Manager (CRM) confirmed this, saying: "We focused on the three main ones: health, education, and youth."

Besides developing its strategic foci, the JV has also established its key approaches or "mechanisms" to community engagement as stated in its sustainability report.

> We target our community investment into areas which are likely to deliver the most sustainable benefit and we employ a number of mechanisms to ensure cost effectiveness, such as leveraging off other organizations in the area, targeting activities which are likely to also have spin-off benefits for the company and involving the community in delivering the benefits, thereby ensuring ongoing ownership.
>
> (Sustainability Report, 2012, p. 19)

The three mechanisms can be classified as (1) partnership with other NGOs and business organisations; (2) business case, that is, benefits to business from their CSR activities (Carroll & Shabana, 2010); and (3) direct partnership with community groups.

Programmes and activities. The JV has undertaken a variety of community activities and programmes over the past seven years to implement its community engagement strategy. Table 7.1 summarises the major community programmes and activities and their delivery mechanisms.

In order to understand why the JV undertakes such community engagement strategies and initiatives, and uses such approaches, we need to examine its motivation.

Community engagement motivations

Institutional drivers

How were these community engagement strategies and initiatives developed? Institutional theory and stakeholder theory are a very powerful means of explaining this move. From an institutional perspective, drivers for community engagement have come from international, national, industry and society levels.

At the international level, the GRI has been widely used in the Australian mining industry. GRI has also issued a mining and metals sector disclosure in addition to its proposed general reporting practice (Global Reporting Initiative, 2013). The JV has incorporated sustainability in its annual reporting and published a sustainability report annually.

At the national and state levels, community engagement is included in regulations governing the Australian mining industry. For example, the Western Australian government requires mining companies to negotiate with the tradition landowners wherever mining is planned in

Table 7.1 The community engagement programmes and activities undertaken by the JV

Practices/activities	Partnership with other NGOs and/or commercial organisations	Based on the company's resources	Partnership with community
Regional health			
RFDS	Donation to RFDS, together with government funding and other business organisations		
Mobile dental clinic	Main sponsor, collaboration with RFDS and two more private enterprises providing services to the mining industry and to three local governments		
Regional general practitioner (GP) flights		Provision of 26 free return flights to Perth a year on its charter flight service	
Community health fund		Contribution to the Community Health Fund	
Education		Provision of 13 apprenticeships, eight tertiary scholarships and eight traineeships to the local community. Total costs so far more than US$450,000	

Table 7.1 (Continued)

Practices/activities	Partnership with other NGOs and/or commercial organisations	Based on the company's resources	Partnership with community
Youth development	Donation to Parkerville Children and Youth Care – a regional NGO – for providing preventive services for children and youth in the region		
Community farm			Donation of free farmland to local community groups as an alternative means of fundraising
Entrepreneurship development – innovation park	Donation of free farmland and provision of initial upfront fund with contractors and work with local NGO to employ indigenous people and develop their business skills		
Targeted employment	Leverage of its procurement power to encourage contractors to employ indigenous people (36)	Direct employment of 16 indigenous people in the company	

order to reach a Native Title Agreement before the exploration license is issued. In Victoria, the community engagement process needs to be followed as required by the Victorian government for a mining operation (Department of Environment and Primary Industries, n.d.).

At the industry level, the national mining industry association, the MCA, has issued its Enduring Value – The Australian Minerals Industry Framework for Sustainable Development (Minerals Council of Australia, 2005) and explicitly states that the industry vision is "[a] valued Australian minerals industry achieving outstanding environmental, social and economic performance".

At the society and community levels, corporate community engagement has become a normal expectation in the mining industry. As the Community Relations Manager states:

> There's a perception out there, especially in the Australian public . . . as soon as a mining company comes to town they have a big . . . wish list which grows and grows. That expectation's Australia-wide.

These institutional pressures or expectations at different levels have driven Australian mining companies to consider community engagement as an indispensable part of operational functions, although such a business function may not yet be regarded as being "core" to them (Kemp & Owen, 2013). Some community engagement practices can be explained by these institutional pressures. For example, engaging with indigenous communities, including employment and training, was part of the company's attempt to reach a Native Title Agreement with the traditional landowners; the Senior Indigenous Affairs Advisor (SIAA) commented as follows:

> Out of that [seven native titles] once again the word gets out there that [the company] are doing some good things with the community and it makes our job easier when we go to try and get approval for – but that's not necessarily the reason we're doing it because we want to be a good neighbor.
>
> (SIAA)

In light of the above, multiple-faceted institutional pressure dictates that mining companies in Australia commit to community engagement. This commitment has become a norm and has provided a framework for mining managers to develop community engagement strategy and initiatives. However, the scope of community engagement and the level of a company's contribution to a specific community can vary within the framework. Such a variation in the scope and contribution level of community engagement can be explained with the stakeholder management theory.

Stakeholder management

Although the institutional pressure mentioned above has influenced the company studied to consider community engagement as necessary for its operations, stakeholder management seems to dictate what community engagement programmes should be developed and how they can be effective and efficient.

Regarding the scope of community groups to be engaged, the linear infrastructure and its operations have determined geographically the community groups with which the company needs to engage. For example, in recognising that the linear infrastructure involves five shires and their potential competition for funding and support, the company has often developed community initiatives that can serve the region, rather than individual shires. This has been stressed by the Community Relations Manager:

> We have a lot of linear infrastructure that goes through these communities. So we have a railway line, a power line, and a water pipeline which goes through a fair bit of freehold land and obviously through these communities and through the shires. So to have their support – because it's freehold land and things like that they've got to give us their support and their approval to build some of this infrastructure through there. So hence you've got to work with them from that point of view. You've got to keep work – make the community happy and want us to be there. It is an expectation that happens within the communities. By doing what we're doing and getting it out there, I think it makes our license to operate and get these approvals through a lot easier.
>
> (CRM)

Other examples of these regionally focused initiatives include the mobile dental clinic, sponsorship of the Royal Flying Doctor Service of Australia and the community farm and innovation park. Having community support, and gaining approval and legitimacy, is the major reason for such community engagement.

Gaining government support and approval is another motivation for the mining company to engage local communities. Government is a powerful stakeholder for business. Although governments at the federal, state and local levels have not exerted direct pressure for the company to engage communities, except in the case of a Native Title Agreement, community engagement activities can help the company obtain government approval due to the democratic nature of the governments, which are elected by the general public. This democratic relationship between government and community has been well recognised by the company's Community Relations Manager, who stated:

> When the government sees that [community engagement activities], that we were doing some of those things, that makes some of our process easier with dealing with government because we can stand up and show some of these things that we're doing working with the communities. Because at the end of the day those are the communities – those people in the community are the electors and they're the ones that are going to go back and complain to the government.
>
> (CRM)

Such a drive to build relationships with local government was reflected by the company's substantial contribution in offering scholarships to a senior high school at a local shire. The shire government has a vision to build the shire into an education hub and already has one technical college and high school located in the shire. The company's contribution to the scholarship has been highly valued by the local government, as one senior official commented:

> They [the company] focus on education, which for [us] we were more than happy because education is a strong wish of ours to make that better and to give more opportunity to the rural students. So they have been absolutely fabulous with regard to scholarships and apprenticeships, training opportunities and to be quite honest we couldn't ask for better.

Conclusion

Two primary motivations for the company to engage communities are institutional pressure and stakeholder management. Institutional pressure from international, national, state and community bodies has made community engagement an indispensable function in the Australian mining industry.

A variety of community engagement programmes and activities have been undertaken by the company based on its analyses of stakeholders as well as internal resources and capabilities. The legitimacy and power of the key stakeholders, coupled with the company's context (operations, resources and locations), have influenced its community engagement strategy and approach. Providing additional health services and education have been the dominant programmes and activities undertaken by the company due to the demand and expectation of local communities and governments.

The community engagement approaches adopted by the company tend to be partnerships with other commercial, public and not-for-profit organisations. The other main approaches are leveraging its contract power with contracting partners based on its resources and strengths. These approaches have made the company's community engagement programmes sustainable. Understanding the local and organisational contexts has been crucial.

Two implications can be suggested based on the conclusions. First, community engagement is context-specific. Understanding local context becomes critical to the development of sustainable and innovative community engagement strategies, approaches, programmes and activities. It is also important for organisations to analyse its internal resources and strengths and participate in business and social networks. All these factors are fundamental for organisations to develop effective community engagement strategy, leverage resources with other business organisations and NGOs for community engagement and, thus, deliver real benefits to communities in a sustainable way.

For a start-up mining company without cash flow, partnering with other organisations could prove a very effective way to deliver benefits to its communities. The real challenge for partnership is how to leverage its resources and strengths, negotiate with potential partners and even take a leading position. Given the social benefit derived from community engagement programmes, a different set of leadership and negotiation skills vis-à-vis potential partners – such as governments, NGOs, special community groups and business organisations – may be required. These skills have historically not been core to mining companies.

Limitations and directions for future research

Two major limitations exist in this chapter. First, this chapter only examined the community engagement strategy and motivations in one Australian mining company. Thus, the findings may be specific to the company examined and may not represent the general community engagement in the mining industry in terms of motivations, practices and approaches adopted. A large sample size can generate a more comprehensive picture of the community engagement in the Australian mining industry. The second limitation, which is relevant to the first one, is that this chapter only investigated the community engagement specific to the Australian mining industry. Thus, the findings of this chapter cannot be generalised to firms in other industries. Future research could focus on a comparative study between the mining and other industries in Australia. The nature of industry can have a considerable impact on the community engagement programmes undertaken by business organisations. An industry-comparative study would enhance our understanding about how industry influences community engagement programmes.

References

Aguinis, H. & Glavas, A. (2012). What we know and don't know about corporate social responsibility: A review and research agenda. *Journal of Management, 38*(4), 932–968.

Banerjee, S. B., Iyer, E. S., & Kashyap, R. K. (2003). Corporate environmentalism: Antecedents and influence of industry type. *Journal of Marketing, 67*(2), 106–122.

Bansal, P. (2005). Evolving sustainably: A longitudinal study of corporate sustainable development. *Strategic Management Journal, 26*(3), 197–218.

Bansal, P. & Roth, K. (2000). Why companies go green: A model of ecological responsiveness. *Academy of Management Journal, 43*(4), 717–736.

Barnett, M. L. (2007). Stakeholder influence capability and variability of financial returns to corporate social responsibility. *Academy of Management. The Academy of Management Review, 32*(3), 794–816.

Barnett, M. L. & Salomon, R. M. (2006). Beyond dichotomy: The curvilinear relationship between social responsibility and financial performance. *Strategic Management Journal, 27*(11), 1101–1122.

Bice, S. (2013). No more sun shades, please: Experiences of corporate social responsibility in remote Australian mining communities. *Rural Society, 22*(2), 138–152.

Bondy, K., Moon, J., & Matten, D. (2012). An institution of corporate social responsibility (CSR) in multi-national corporations (MNCs): Form and implications. *Journal of Business Ethics, 111*(2), 281–299. doi:10.1007/s10551-012-1208-7

Bowen, F., Newenham-Kahindi, A., & Herremans, I. (2010). When suits meet roots: The antecedents and consequences of community engagement strategy. *Journal of Business Ethics, 95*(2), 297–318. doi:10.1007/s10551-009-0360-1

Bowen, H. R. (1953). *Social responsibilities of the businessman*. New York: Harper & Row.

Brammer, S. & Millington, A. (2004). The development of corporate charitable contributions in the UK: A stakeholder analysis. *The Journal of Management Studies, 41*(8), 1411–1434.

Carroll, A. B. (1979). A three-dimensional conceptual model of corporate performance. *Academy of Management Review, 4*(4), 497–505.

Carroll, A. B. & Shabana, K. (2010). The business case for corporate social responsibility: A review of concepts, research and practice. *International Journal of Management Reviews, 12*(1), 85–105.

Clarke, T. (2007). The materiality of sustainability: Corporate social and environmental responsibility as instruments of strategic change? In S. Benn & D. Dunphy (Eds.), *Corporate governance and sustainability: Challenges for theory and practice* (pp. 219–251). London: Routledge.

Clarkson, M. E. (1995). A stakeholder framework for analyzing and evaluating corporate social performance. *Academy of Management Review, 20*(1), 92–117. doi:10.5465/amr.1995.9503271994

CNOOC. (2014). Key CSR Practices by CNOOC in Australia. In Z. Ma (Ed.), *How to invest in Australia* (pp. 204–207). Canberra: China Commerce and Trade Press.

Darnall, N., Henriques, I., & Sadorsky, P. (2010). Adopting proactive environmental strategy: The influence of stakeholders and firm size. *The Journal of Management Studies, 47*(6), 1072–1094.

Department of Environment and Primary Industries. (n.d.). What is community engagement? Retrieved 1 December 2013, from http://www.dse.vic.gov.au/effective-engagement/introduction-to-engagement/what-is-community-engagement

DiMaggio, P. J. & Powell, W. W. (1983). The iron cage revisited: Institutional isomorphism and collective rationality in organizational fields. *American Sociological Review, 48*(2), 147–160.

Donaldson, T. & Preston, L. E. (1995). The stakeholder theory of the corporation: Concepts, evidence, and implications. *Academy of Management Review, 20*(1), 65–91. doi:10.5465/amr.1995.9503271992

Drysdale, P. & Findley, C. (2008). Chinese foreign direct investment in Australia: Policy issues for the resource sector. Canberra. Retrieved 31 January 2012, from http://www.eastasiaforum.org/wp-content/uploads/2008/09/drysdale_and_findlay_chinese_fdi_300808.pdf.

Ellem, B. (2003). Contested communities: Geo-histories of unionism. *Journal of Organizational Change Management, 21*(4), 433–450.

Eweje, G. (2006). The role of MNEs in community development initiatives in developing countries. *Business & Society, 45*(2), 93–129.

Freeman, R. E. (1984). *Strategic management: A stakeholder approach.* Boston: Pitman Publishing.

Friedland, R. & Alford, R. R. (1991). Bring society back in: Symbols, practices and institutional contradiction. In W. W. Powell & P. J. DiMaggio (Eds.), *The new institutionalism in organizational analysis* (pp. 232–263). Chicago: The University of Chicago Press.

Gibbert, M. & Ruigrok, W. (2010). The "What" and "How" of case study rigor: Three strategies based on published work. *Organizational Research Methods, 13*, 710–737. doi:10.1177/1094428109351319

Gifford, B., Kestler, A., & Anand, S. (2010). Building local legitimacy into corporate social responsibility: Gold mining firms in developing nations. *Journal of World Business, 45*(3), 304–311. doi:10.1016/j.jwb.2009.09.007

Global Reporting Initiative. (2013). G4 sector disclosure – Mining and metals. Retrieved 1 May 2014, from https://www.globalreporting.org/resourcelibrary/GRI-G4-Mining-and-Metals-Sector-Disclosures.pdf

Gugler, P. & Shi, J. Y. (2009). Corporate social responsibility for developing country multinational corporations: Lost War in Pertaining Global Competitiveness? *Journal of Business Ethics, 87*, 3–24. doi:10.1007/s10551-008-9801-5

Hart, S. L. (1995). A natural-resource-based view of the firm. *Academy of Management. The Academy of Management Review, 20*(4), 986–1014.

Henriques, I. & Sadorsky, P. (1996). The determinants of an environmentally responsive firm: An empirical approach. *Journal of Environmental Economics and Management, 30*(3), 381–395.

Huang, X., & Austin, I. (2011). *Chinese investment in Australia: Unique insights from the mining industry*. New York: Palgrave Macmillan.

Huang, X. & Staples, W. (2014). Community engagement by Chinese firms in Australia: Practices and benefits. Melbourne: RMIT University.

Idemudia, U. (2009). Assessing corporate-community involvement strategies in the Nigerian oil industry: An empirical analysis. *Resources Policy, 34*(3), 133–141. doi:10.1016/j.resourpol.2009.01.002

Jiang, R. J. & Bansal, P. (2003). Seeing the need for ISO 14001. *The Journal of Management Studies, 40*(4), 1047–1067.

Jones, M., Marshall, S., & Mitchell, R. (2007). Corporate social responsibility and the management of labour in two Australian mining industry companies. *Corporate Governance, 15*(1), 57–67.

Kemp, D. & Owen, J. R. (2013). Community relations and mining: Core to business but not "core business". *Resources Policy, 38*(4), 523–531. doi:10.1016/j.resourpol.2013.08.003

Kemp, D., Owen, J. R., & van de Graaff, S. (2012). Corporate social responsibility, mining and "audit culture". *Journal of Cleaner Production, 24*(0), 1–10. doi:10.1016/j.jclepro.2011.11.002

Kostova, T., Roth, K., & Dacin, M. T. (2008). Institutional theory in the study of multinational corporations: A critique and new directions. *Academy of Management. The Academy of Management Review, 33*(4), 994–1006.

KPMG. (2011). International survey of corporate responsibility reporting. Retrieved 3 January 2012, from http://www.kpmg.com/Global/en/IssuesAndInsights/ArticlesPublications/corporate-responsibility/Pages/2011-survey.aspx

Lee, M.-d. P. (2008). A review of the theories of corporate social responsibility: Its evolutionary path and the road ahead. *International Journal of Management Reviews, 10*(1), 53–73.

Luo, Y., Shenkar, O., & Nyaw, M. -K. (2002). Mitigating liabilities of foreignness: Defensive versus offensive approaches. *Journal of International Management, 8*(3), 283–300.

Marquis, C. & Battilana, J. (2009). Acting globally but thinking locally? The enduring influence of local communities on organizations. *Research in Organizational Behavior, 29*, 283–302. doi:10.1016/j.riob.2009.06.001

Miles, M. B. & Huberman, A. M. (1994). *Qualitative data analysis*. London: Sage Publications.

Minerals Council of Australia. (2005). Enduring value – The Australian minerals industry framework for sustainable development. Retrieved 1 January 2014, from http://www.minerals.org.au/focus/sustainable_development/enduring_value

Ministry of Commerce. (2014). *2013 Statistical bulletin of China's outward foreign direct investment*. Beijing: China Statistics Press.

Mitchell, R. K., Agle, B. R., & Wood, D. J. (1997). Toward a theory of stakeholder identification and salience: Defining the principle of who and what really counts. *Academy of Management Review, 22*(4), 853–886. doi:10.5465/amr.1997.9711022105

Muthuri, J. N., Chapple, W., & Moon, J. (2009). An integrated approach to implementing "community participation" in Corporate Community Involvement: Lessons from Magadi Soda Company in Kenya. *Journal of Business Ethics, 85*(2), 431–444. doi:10.1007/s10551-008-9739-7

Muthuri, J. N., Moon, J., & Idemudia, U. (2012). Corporate innovation and sustainable community development in developing countries. *Business & Society, 51*(3), 355–381. doi:10.1177/0007650312446441

North, D. (1991). Institutions. *Journal of Economic Perspectives, 5*(1), 97–112.

North, D. (1992). Institutions, ideology, and economic performance. *CATO Journal, 11*(3), 477.

Orlitzky, M., Frank, L. S., & Sara, L. R. (2003). Corporate social and financial performance: A meta-analysis. *Organization Studies, 24*(3), 403–441.

Owen, J. R. & Kemp, D. (2012). Assets, capitals, and resources: Frameworks for corporate community development in mining. *Business & Society, 51*(3), 382–408. doi:10.1177/0007650312446803

Owen, J. R. & Kemp, D. (2013). Social licence and mining: A critical perspective. *Resources Policy, 38*(1), 29–35. doi:10.1016/j.resourpol.2012.06.016

Owen, J. R., & Kemp, D. (2014). Mining and community relations: Mapping the internal dimensions of practice. *The Extractive Industries and Society,1*(1), doi:10.1016/j.exis.2013.12.004

Parmar, B. L., Freeman, R. E., Harrison, J. S., Wicks, A. C., Purnell, L., & de Colle, S. (2010). Stakeholder theory: The state of the art. *Academy of Management Annals, 4*(1), 403–445. doi:10.1080/19416520.2010.495581

Peng, M. W. (2003). Institutional transitions and strategic choices. *Academy of Management. The Academy of Management Review, 28*(2), 275–296.

Porter, M. E., & Kramer, M. R. (2011). Creating shared value. *Harvard Business Review, 89*(1/2), 62–77.

Post, J. E., Preston, L. E., & Sachs, S. (2002). *Redefining the corporation: Stakeholder management and organizational wealth*. Stanford, CA: Stanford University Press.

Rowley, T. J. (1997). Moving beyond dyadic ties: A network theory of stakeholder influences. *Academy of Management Review, 22*(4), 887–910. doi:10.5465/amr.1997.9711022107

Sauser Jr, W. (2005). Ethics in business: Answering the call. *Journal of Business Ethics, 58*(4), 345–357. doi:10.1007/s10551-004-5715-z

Schmidt, A. H. & Freeman, S. J. (2000). Corporate social performance and attractiveness as an employer to different job seeing populations. *Journal of Business Ethics, 28*(3), 243–253.

Scott, R. (1987). The adolescence of institutional theory. *Administrative Science Quarterly, 32,* 493–511.

Scott, W. R. (1995). *Institutions and organizations*. Thousand Oaks: SAGE.

Sharma, S. (2000). Managerial interpretations and organizational context as predictors of corporate choice of environmental strategy. *Academy of Management Journal, 43*(4), 681–697.

Sharma, S., Aragón-Correa, J. A., & Rueda-Manzanares, A. (2007). The contingent influence of organizational capabilities on proactive environmental strategy in the service sector: An analysis of North American and European ski resorts. *Canadian Journal of Administrative Sciences, 24*(4), 268–283.

Sharma, S., Pablo, A. L., & Vredenburg, H. (1999). Corporate environmental responsiveness strategies: The importance of issue interpretation and organizational context. *The Journal of Applied Behavioral Science, 35*(1), 87–108.

Slack, K. (2011). Mission impossible?: Adopting a CSR-based business model for extractive industries in developing countries. *Resources Policy, 37*(2), 179–184.

State-owned Assets Supervision and Administration Commission (SASAC). (2008). Guidelines to the state-owned enterprises directly under the central government on fulfilling corporate social responsibilities. Retrieved 1 January 2014, from http://en.sasac.gov.cn/n1408035/c1477196/content.html

Suchman, M. C. (1995). Managing legitimacy: Strategic and institutional approaches. *Academy of Management Journal, 20*(3), 571–610.

Trevino, L. K. & Nelson, K. A. (2014). *Managing business ethics: Straight talk about how to do it right* (6th ed.). San Francisco, CA: Wiley.

Turban, D. B. & Greening, D. W. (1997). Corporate social performance and organizational attractiveness to prospective employees. *Academy of Management Journal, 40*(3), 658–672.

Vermander, B. (2014). *Corporate social responsibility in China: A vision, an assessment and a blueprint.* Singapore: World Scientific Publishing.

Waddock, S. A., & Samuel, B. G. (1997). The corporate social performance-financial performance link. *Strategic Management Journal, 18*(4), 303–319.

Wartick, S. L. & Cochran, P. L. (1985). The evolution of the corporate social performance model. *Academy of Management Review, 10*(4), 758–769. doi:10.5465/amr.1985.4279099

Wokutch, R. E. (1990). Corporate social responsibility Japanese style. *Executive (19389779), 4*(2), 56–74. doi:10.5465/ame.1990.4274797

Wright, M., Filatotchev, I., Hoskisson, R. E., & Peng, M. W. (2005). Strategy research in emerging economies: Challenging the conventional wisdom. *The Journal of Management Studies, 42*(1), 1–33.

Yin, R. K. (2003). *Case study research: Design and methods* (3rd ed.). Thousand Oaks, CA: Sage.

Zadek, S., Forstater, M., & Yu, K. (2012). Corporate responsibility and sustainable economic development in China: Implications for business. Retrieved 15 March 2012, from http://www.uschamber.com/sites/default/files/international/files/17296_China%20Corp%20Social%20Responsibility_Opt.pdf

8
Conclusion: Ongoing Challenges for Chinese OFDI and MNCs Operating Abroad

Introduction

After nearly four decades of economic reform, China has accumulated huge foreign reserves and has become the second largest economy in the world. The current national economic policies not only promote further economic integration in the global economy by attracting foreign capital, management expertise and "know-how", but also encourage Chinese companies, in particular large SOEs, to go abroad and carry out OFDI in countries with advanced technology, rich natural resources and potential markets in order to secure technological and raw material supplies as well as international market expansion for the future economic development in China. Therefore, we have been able to observe a large number of Chinese MNCs investing in other developed and developing countries in recent years; Australia being one of them.

This book is the first of its kind to comprehensively document and analyse the pattern, strategy and historical background of Chinese OFDI and MNCs operating in Australia. Key issues relevant to the MNCs' international operations have been included, such as the entry strategy, investment timing, the stage of project invested and the liabilities of foreignness and newness. These factors influence the Chinese MNC's overall business performance as latecomers. In addition, this book provides a number of detailed case studies in order to address the fundamental issues of international business management, such as cross-border acquisition and managing post-transaction integration, corporate governance and control of overseas subsidiaries, cross-cultural management and international HRM, and local community engagement. From the research findings presented in these case studies, we can see that there are many ongoing challenges for Chinese OFDI and

MNCs operating abroad. In this concluding chapter, we highlight these challenges and discuss some implications for improving future development, in the hope that the relevant decision-makers and business leaders both in China and abroad can learn from these lessons and be more successful in future OFDI and MNC operations.

Ongoing challenges for Chinese OFDI and MNCs operating abroad

In recent years, increasingly more Chinese capital has been invested in Australia. Large SOEs as well as other relatively small-scale private companies have been involved in such OFDI activities. The overall performance of Chinese OFDI in Australia, with some exceptions, is unsatisfactory due to the entry strategy, such as investment timing, the stage of project invested and the liability of foreignness and newness. In our studies, the majority of Chinese investments in the Australian mining industry have underperformed. There are also clear patterns between performance, the timing of entry and the target company's developmental stage. It is clear to us now that investments generally should not be made at the peak of economic cycles, as assets are often expensive and overvalued. In addition, the company's developmental stage is also crucial to investment performance: investments in target companies at the production stage can perform better than those made at the exploration and developmental stages.

Furthermore, a range of crucial information is necessary for sound investment decisions by the Chinese government and senior leadership of Chinese MNCs, such as the host government's approval procedures and regulations, the target company's background and management, and environmental protection and community engagement. These issues often challenge Chinese MNCs in developed countries like Australia. The reason for this is that many Chinese investors do not have much experience in international operations largely because of their long history of isolation from the global market. OFDI by Chinese firms is a relatively new phenomenon and has only gained momentum since 2003. Players in such a phenomenon are labelled the "latecomers" who suffer "liability of foreignness". The consequence may be poor investment performance. The disadvantage of being "latecomers" is that Chinese investors face difficulties in competing with other well-established MNCs with world-class capacity. Therefore, lack of international business operation experience and insufficient information about the host country's political, legal, social/cultural and environmental policies and norms could jeopardise the success rating of Chinese OFDI and MNCs' operations abroad.

Another obvious problem for Chinese investors is that most of them have a strong desire to control the companies in which they invest. Due to cultural, historical and administrative legacies, Chinese investors find it difficult to trust an outsider to manage their businesses and investments. Moreover, corporate governance of Chinese MNCs is still weak, and these factors can also influence Chinese MNCs performing well in overseas investments.

Another challenging issue is related to cross-border acquisition and managing post-transaction integration. In fact, in the last decade, Chinese MNCs have adopted cross-border acquisition as their major approach in their OFDI. There are some advantages of adopting the OFDI method. For example, acquiring a well-managed company in a developed country is a quick way for Chinese MNCs to catch up with their international competitors and further their international expansion. Meanwhile, such an acquisition is a very effective way for Chinese MNCs to overcome their "liability of foreignness" and their comparative disadvantage to operating in developed countries.

Another advantage of using this approach is that Chinese MNCs can explore the cooperation and coordination of value chain activities to reduce procurement costs or increase sales by leveraging the acquiring company's expertise and capability. In addition, acquiring a relatively large Australian company and retaining all senior managers has a major potential for reverse knowledge transfers, both technically and managerially. Knowledge and skills at organisational, functional and task levels can be potential targets to be transferred between subsidiaries and headquarters. Relying on the local senior managers to play the leadership role in Chinese MNCs is also extremely important to the success of post-acquisition integration.

The issue of leadership in Chinese MNCs leads us to consider the relevant cross-cultural management and international HRM policies and practices. Many Chinese companies operating their business overseas face the challenges of cross-cultural management. Adequate international HRM policies and practices can enable them to overcome these problems. Our case studies demonstrated that a number of factors influence Chinese MNCs' decisions about whether to adopt appropriate international HRM policy and practices. The overall picture is that the quality and capability of Chinese managers are inadequate in terms of cross-cultural communication and management and demonstrate an unfamiliarity with local laws and regulations, a lack of autonomy to make key decisions in the financial and HR areas and an ethnocentric approach to carrying out their own management styles and attitudes in their overseas business operations (e.g. face culture and relationships (*guanxi*) are more important than "rule of law").

The last but not insignificant issue is the Chinese MNCs' engagement with local communities. The demand is increasing from the general public, customers and communities for MNCs to take responsibility in both social and ecological areas. Institutional investors and major MNCs are increasingly expected to take corporate sustainability issues seriously. Community engagement is a vital part of CSR and has become an essential element of operations in the development of sustainable business overseas. So far, a variety of community engagement programmes and activities have been undertaken by Chinese MNCs, such as providing additional health services and education to meet the expectation of local communities and governments. In addition, the community engagement approaches adopted by Chinese MNCs tend to be partnerships with other commercial, public and not-for-profit organisations. These approaches have rendered the Chinese MNCs' community engagement programmes sustainable. Through these engagements, the Chinese MNCs can have a better understanding of the local political, legal, social/cultural and environmental policies and norms.

Conclusion

In Table 8.1, we summarise the key problems and challenges existing for the current Chinese MNCs operating abroad and possible options for better policies and practices that are crucial for future successful overseas business operations. Based on these considerations, we elaborate some alternatives towards future improvements.

The overwhelming impression gained in our studies of Chinese OFDI and MNCs' operations abroad indicates that China has accumulated a large volume of foreign reserves in recent decades, and it is time for China to use them through OFDI to secure supplies of natural resources, expand market opportunities and obtain advanced technology and management "know-how". However, the lesson we learnt in compiling this book is that cash alone is not sufficient for successful OFDI. The most challenging issue is the lack of adequate overseas business operation experience of the majority of Chinese MNCs, regardless of whether the experience is related to OFDI decision-making and entry strategies or to corporate governance, leadership, cross-cultural management/international HRM and community engagement. We have

Table 8.1 The existing problems and challenges for Chinese MNCs and future improvements for policies and practices

Issues	Problems and challenges	Future improvements
Strategic entry: • timing • stage of project invested • percentage of ownership	Investments made at the peak of economic cycles are often expensive and overvalued 'Greenfield' operations have more difficulties, particularly at the stage of exploration and development stage Ownership issue is not a fundamental problem, but other factors such as the quality of the board and management team, cultural clash and management turnover play a crucial role in influencing financial performance	Avoiding investment during the peak of economic cycles Avoiding administrative intervention from the governments and senior officials Investment decision-making cannot be influenced very much by Chinese national, social and political considerations, but should be based on business rationale Comprehensive view of the entire business/product life cycle with better understanding about the key uncertainties involved in each stage, and thus better managing investment risks and improving investment performance Fully utilising local talents in daily business operation and organisational management regardless of whether Chinese investors are majority or minority shareholders
"Liability of foreignness" and newness	Lack of information about the business investment and operation process due to the "liability of foreignness" can contribute to the poor performance of investment As "newcomers" or "latecomers", Chinese MNCs have many difficulties to compete with other well-established MNCs and developing sound businesses overseas	Being better prepared before making investment decisions by gathering information and conducting due diligence Investing as a minority shareholder in a listed company can be a viable strategy to avoid the disadvantage of being a "newcomer" with lower risk while developing business through ongoing learning from local partners, and, in this sense, joint ventures with successful local companies can be an effective investment approach

Post-acquisition integration	Corporate control is one of the most important challenging issues for post-acquisition integration Both control structure and control approach are problematic but important for successful outcomes The challenge of knowledge transfer between headquarters and subsidiaries The challenge of leadership in managing the post-acquisition integration	Establishing a board and adopting a strategic control approach An effective board can buffer the potential formal or informal interferences from the Chinese government and other leaders who are not familiar with the business situation Adopting an integration approach combined with not sending expatriates to executive positions and fully using existing local management teams in order to learn and transfer expertise back to headquarters Local talent leadership in management is crucial for the overseas business operation
Corporate control by Chinese MNCs over their subsidiaries	Applying corporate control logics in the parent company to foreign subsidiaries, with the emphasis on manager control rather than corporate governance Appointing insiders rather than outsiders as CEO/MD to subsidiaries Lack of experienced executives and middle managers, coupled with potential language problems Control logic relies heavily on ownership and personal control and has been considered as counterproductive	Adopting modern corporate governance systems in international business operations Relying on competency but not a *guanxi*-based approach to select CEO/MD with the intention of fully utilising local talent leadership Implementing a new institutional logic with consideration of the material practices, cultural symbols and the cross-level effects from national, organisational, social and individual perspectives

Table 8.1 (Continued)

Issues	Problems and challenges	Future improvements
Cross-cultural management and international HRM	The quality and capability of Chinese managers are inadequate Lack of cross-cultural communication and management capabilities Unfamiliar with local laws and regulations Lack of autonomy to make key decisions in the financial and HR areas Having an ethnocentric approach in management styles and attitudes in overseas operations	Establishing a global talent pool for reserving, developing and allocating capable leaders for overseas business operations Pre-departure cross-cultural training on communications and management in cross-cultural environments Ongoing learning and training for Chinese expatriates with a wide range of issues, including laws and regulations, economic, social and cultural knowledge, HR and labour management issues Arranging partnership and mentor relationship between Chinese expatriates and local managers/employees in order to facilitating cross-cultural understanding and friendship
Engagement with local communities	Increasing demand from the general public, customers and communities for MNCs to take CSR seriously Challenges for business decisions not only with regard to making profit, but also considering the well-being of communities, environments and the whole society	Establishing KPI related to community engagement and MNCs' CSR Actively participating in community activities and providing support Regular meetings with community leaders and understanding community needs Setting up business goals with consideration of triple bottom lines, namely economic, social and environmental consequences

observed a number of fundamental problems in the timing and entry mode of OFDI; ethnocentric governance systems; shortage of capable leaders with international business experience and cross-cultural management competence; and unfamiliarity with local political, legal, social/cultural and environmental policies and norms.

These problems have led us to consider possible solutions by adopting new policies and practices for improving future OFDI activities. The first step is to build adequate human capital by obtaining and developing a globally oriented talent pool and fully utilising capable leadership not only within Chinese MNCs' systems but also in wider business communities globally. This will challenge the current practices of using "insiders" only. The intervention of government officials and senior leadership within MNCs in appointing leadership at the subsidiary level should be limited and replaced by international professional practices of appointing leadership through proper HR channels, formalities and procedures. Screening the capability of leadership, rather than utilising MNCs' internal organisational hierarchy and *guanxi* to select leaders for managing overseas investment operations, could be a vital alternative to current practices and a fundamental element for future success.

In order to make investment decisions and select entry modes more appropriately, the Chinese MNCs should initially establish an "Investment Advisory Board" that includes host country members with local legal, accounting, finance, environment and cross-cultural management expertise. The Board members could provide sound investment advice before making the final investment decision and prevent serious decision-making mistakes, as was seen in our case studies. In addition, an increasing number of capable overseas Chinese professionals are available in the host countries (i.e. Australia, Canada, Europe, New Zealand, the UK and the US) as a result of more than three decades of "open-door" policy on "overseas education of Chinese students" implemented by the Chinese central government. Nowadays, a wide variety of well-trained Chinese professionals is working in major Western countries' education sectors and industries with a wide range of connections. The Chinese government and Chinese MNCs should establish a data pool in this regard and develop connections with these Chinese professionals abroad, identifying ways to fully utilise their "know-how" and social networks in these well-developed markets.

Another important area for improvement is providing full autonomy to Chinese subsidiaries in decision-making. Sound and successful OFDI decisions require fewer Chinese government interventions (as a political task) and more leadership, financial and HR support from the headquarters of Chinese MNCs. Currently, the majority of Chinese subsidiaries do not have sufficient managerial autonomy in the areas of finance and HR management. At the subsidiary level, limited capacity to do things with insufficient budgets and limited power for appointing senior managers, recruiting and promoting local employees, training both Chinese expatriates and local managers/employees, and insufficient support for local engagement (i.e. lack of funding) jeopardise the opportunity of developing a fully functioned and capable subsidiary. A "daughter-in-law" phenomenon in these overseas Chinese subsidiaries under the sever control of their parent companies in China is widespread and is not healthy for building successful Chinese MNCs in the long run.

Finally, we hope this book can provide some insight into and knowledge of Chinese OFDI and MNCs operating overseas. By understanding the problems and challenges, we can develop some alternative policies and practices to overcome these problems and achieve a better performance outcome. At the end of the book, we, the authors, call upon the Chinese government and senior leaders among the Chinese MNCs to take appropriate action to correct the wrongdoings of the past and to remember that there are many Chinese or foreign professional people who are willing to facilitate Chinese OFDI and MNCs to achieve long-term sustainable development globally.

Index

CPSIA information can be obtained at www.ICGtesting.com
Printed in the USA
BVOW06*0217090616

451363BV00006B/21/P

9 781137 394583